Process Development Division

Core Programming Topic at the 2012 AIChE Spring Meeting and 8th Global Congress on Process Safety

Houston, Texas, USA
1-5 April 2012

ISBN: 978-1-62276-557-7

Printed from e-media with permission by:

Curran Associates, Inc.
57 Morehouse Lane
Red Hook, NY 12571

Some format issues inherent in the e-media version may also appear in this print version.

Printed by Curran Associates, Inc. (2012)

For permission requests, please contact AIChE
at the address below.

AIChE
3 Park Avenue
New York, NY 10016-5991

Phone: (203) 702-7660
Fax: (203) 775-5177

www.aiche.org

Additional copies of this publication are available from:

Curran Associates, Inc.
57 Morehouse Lane
Red Hook, NY 12571 USA
Phone: 845-758-0400
Fax: 845-758-2634
Email: curran@proceedings.com
Web: www.proceedings.com

TABLE OF CONTENTS

(4a) Equations of Change: How Patent Law Has Changed, and What Changes Are to Come

Jeffrey L. Wendt, The Wendt Firm PC, The Woodlands, TX

Patent law, particularly in the United States, has undergone many changes in the last 20 years, both in the courts and in the patent statute. My focus will be on what those changes have been, what changes are likely to come, and how these changes impact practicing chemical engineers. For example, 20 years ago the so-called "doctrine of equivalents" was widely used to ensnare infringers of patents. Now, the doctrine is almost an afterthought. My thoughts will touch on whether this is good or bad. Another change has been the term of patents. Before 1995, an issued patent had a term of 17 years from issue, even if the patent application had been filed many years before. Now, patents are granted for 20 years from the earliest filing date claimed as priority. Finally, I will touch on how a change to a "first to file" patent regime will make life different for engineer/inventors than the present "first to invent" system, particularly for small entities.

(4b) What Is Prior Art in the "First-to-File" Patent System Under the America Invents Act of 2011

Mark Rygiel, Sterne Kessler, Washington, DC and Anbar Khal, Sterne, Kessler, Goldstein & Fox PLLC, Washington, DC

On September 16, 2011, the Leahy-Smith America Invents Act was signed into law by President Obama. This law brings sweeping change to the current U.S. patent system, including transforming the current first-to-invent system into a first-inventor-to-file system. The first-to-file provisions of the Act take effect on March 16, 2013. These provisions substantially overhaul what acts are considered to be "prior art" and what acts are excepted. This presentation will provide an in-depth review of the new scope of prior art under the America Invents Act of 2011, focusing on what forms of disclosure prior to the effective filing date of a patent application will prevent the applicant from obtaining a patent in the United States. Case studies will be presented to illustrate the new prior art provisions, their exceptions, and how the new provisions compare to the existing prior art standards. The presentation will touch on how these new prior art provisions compare to the prior art rules of other developed countries having first-to-file systems.

Avoiding Common Mistakes in Licensing Agreements and Other Technology Agreements

By: Anthony P. Venturino, Esq. , Partner of Novak, Druce + Quigg, LLP Washington, DC

AIChE Spring Meeting - April 2, 2012 Houston, TX

The views presented here are those of the author and not attributable to his firm or its clients.

I. Technology agreements Can't live with them, can't live without them

Whether you consider technology agreements a help or a hindrance we need them to control the flow of technology. They set the ground rules so everyone involved knows how to handle technology being transferred and/or invented so there are no unwelcome surprises.

Thus, chemical engineers and managers need to know enough to know when they need an agreement or, how to perform if they are operating under an agreement.

In the course of performing their jobs chemical engineers and managers encounter many kinds of agreements having technology or other forms of intellectual property as a critical component. A few such agreements are as follows:

- Confidentiality Agreements
- Vendor Supplier Agreements
- Joint Development/Commercialization Agreements
- Exclusive Licenses / Non-exclusive Licenses
- Co-promotion Agreements
- Employee Agreements
- Settlement Agreements
- Joint Venture Agreements
- Contract R&D Agreements
- Consulting Agreements

II. Things can go wrong before, during and after the agreement

Here are some examples of where things can go wrong:

- Evaluation of the technology
- Selection of partners
- Conduct of the negotiations
- The agreement itself
- Implementing and servicing the agreement

Classic Licensee mistakes include:

- Royalties paid when licensed patent had expired
- Payments made after royalty terms had been satisfied
- Payments made under a trademark license when the trademark is not used

- Patents covered in a license no longer cover products that are currently made, yet royalty is paid against the patent coverage
- Royalties paid at the wrong rate, i.e. one higher or lower than provided Source:

Licensing – A Strategy for Profits, Edward P. White (1990)

Licensors also make mistakes. Here are some classics:

- Failure to take action if royalty payments cease when they should not, e.g., prior to expiration of the patents.
- Failure to monitor royalty payments, e.g. sudden reductions or failure to meet minimum royalties that could trigger a change from exclusive to nonexclusive or termination of the licenses.
- Failure to stay in technical contact with the licensee, with the result of licensee not interested in paying royalties, in maintaining quality, or keeping good records.

Licensing – A Strategy for Profits, Edward P. White (1990)

Before doing the deal.

- Know your business goals, strengths and weaknesses
- Know the available protection for the technology
- Have a realistic idea of the value of the technology
- Be open to creative technology agreements
- Do not let paranoia get in the way of a good deal
- Watch out for foreseeable legal and technical problems
- Admit when you are in over your head and get help
- Admit when you are in over your head and get help from a lawyer or other licensing professional (did I just say that?)

III. Avoiding Common Traps

A. Define ownership and control of joint inventions of improvements

Jointly developed invention often occur when smart people from different companies work together, for example while performing a joint R & D agreement or a variety of other agreements between companies. Joint inventions are jointly owned by US law unless there is a contract stating otherwise. This can cause problems.

If possible, have one party own the invention and, if necessary, grant the other party an exclusive field of use license. The following clause does this:

> In the event a joint invention arises during performance of this Agreement Company A owns the joint invention and controls patent prosecution and Company B gets an exclusive field of use license to the joint invention in the field of _______________.

B. Maintain good records of what the agreement covers or when it expires

The company, its legal department or a law firm should docket the agreement in a database. The docketing record should include: the type of agreement, the parties, the technology covered by the agreement, the agreement deadlines, the expiration date, the contact person and backup contact person, and have a miscellaneous portion for anything unusual.

Preferably an institutional person, for example, a docket clerk or contracts administrator, should be responsible for tracking notices in the docketing system and then relay the information to the business people. The more automatic you can make the timely conveying of such information, the better.

C. Do not accidentally create an agreement

The first time you negotiate with a party, especially if the other party sees this agreement as a one shot deal, be careful not to make records during the negotiations which unexpectedly bind you. Do not let an agreement to agree inadvertently become an agreement.

Summaries of the negotiations signed by both parties can lead to binding agreements even when a clause that might be important to you is missing. Let's take a hypothetical scenario where two parties bring their business people and lawyers to a meeting to negotiate a potential IP license. Mid-afternoon the attorney for the IP owner (potential licensor) had to leave to catch a plane. The potential licensor's business person (let's call him "Bill") had a later plane. The potential licensee kindly offered to take Bill to the airport later so he would not have to waste time in a boring airport. While Bill was by himself with the potential licensee's attorney and business people one of the business people suggested they draft and sign a summary of the day's discussions. They did this and then Bill caught his plane. The next day Bill's company was asked to sign an Assignment document to reflect the assignment they agreed to yesterday. Bill's response was "What assignment?" After a lawsuit the parties settled. The moral of the story is train the business people not to sign anything.

A deal can also inadvertently arise if the parties act like there is an agreement can create an agreement. This sometimes arises where the parties are in a rush to do joint development. While the deal is only a draft they start working together. Sometimes the unsigned draft becomes the *de facto* agreement.

The opposite can also happen. Sometimes you think you have an agreement but in reality you do not, or at least not all the clauses apply.

Take the case of *DataTreasury v. Wells Fargo,* 522 F.3d 1368 (Fed. Cir. 2008). In 2003 WMR e-Pin licensed a patent to Wells Fargo Services Corp. a subsidiary of Wells Fargo & Co. The license included an arbitration clause. In 2006 WMR e-Pin assigned 4 patents to DataTreasury. At least one of the 4 patents was a patent of the license. Afterwards DataTreasury sued Wells Fargo & Co. for patent infringement in court. Wells Fargo & Co. moved to dismiss or stay pending arbitration. Under 5th Circuit law, as applied by the

Federal Circuit, "to be enforceable, an arbitration clause must be in writing and signed by the party invoking it. *DataTreasury* at 1373, quoting *Westmorland v. Sadoux*, 299 F.3d 462, 465 (5th Cir. 2002). Neither Wells Fargo & Co. nor DataTreasury signed the original patent license. Thus, the court held these parties were not bound by the arbitration clause.

D. Remember important clauses

Clause checklists are important tools when reviewing a license to see if all the necessary clauses are present. However, use the right checklist for the type of agreement and the type of IP. Do not assume the clauses in your standard patent license belong in your new trademark license and *vice versa*. Unlike patents which protect the patent holder, trademarks protect the public by assuring the public a good labeled with a particular trademark will have the characteristics they expect. For example, a PIRELLI brand tire has certain handling properties. Trademark licenses without quality safeguards are termed naked licenses and invalidate the trademark rights.

A short checklist is provided in TABLE 1:

TABLE 1
Introductory Clauses
Definitions
Grant of Rights
Consideration/Compensation
Diligence
Product Quality Control
Advertising
3rd Party Infringement Suits
3rd Party Product Liability Suit
Licensor/ Licensee Suits
Termination
Continuing Obligations

A detailed set of clause checklists for a licensing agreement is available upon request from the author.

E. Maintain control of the technology after making the deal

Proprietary technology is a competitive advantage. If the licensor loses control of the technology he loses control of the competitive advantage. Thus, the licensor should determine the level of control he is willing to lose. Maybe the licensor thinks he can permit the licensee to sublicense because he trusts the judgment of licensee's president. What if the president leaves? Thus, the agreement preferably precludes sublicensing or assigning to competitors.

Another common way to lose control is for the licensee to be bought by the licensor's competitor. The Agreement should specify it may be assigned by ONE of the following:

1. Anyone;
2. Goes with the business;
3. Goes with the company;
4. Licensor maintains right of first refusal;
5. Licensor maintains right of first refusal which will not be unreasonably withheld;
6. No one.

F. Promote good communication after making the deal

Problems arise from misunderstandings, lack of thoroughness, carelessness, or failure to administer the agreement at all, i.e., no one in authority is responsible for administration.

The lawyer can structure the agreement to include mechanisms for communication between the licensor and licensee so little problems and misunderstandings do not become big ones. You cannot force the business people to communicate but the mechanisms provide fertile soil for communication. For example, the agreement should include a contact person and back-up contact person. The agreement can also mandate times for periodic meetings.

G. You need the equivalent of a pre-nuptial agreement

Agreements, like romance, can end badly. As with a newlywed couple, the enthusiastic partners to a new business relationship think nothing will go wrong. They can and they do.

Include termination provisions in the agreement to give each party a reasonable way to terminate the agreement. This includes planning for unexpected events, for example, bankruptcy or purchase of the other party. This also includes providing ways to terminate or renegotiate if the economics of the deal no longer make the current agreement viable.

IV. Organize A Licensing Department – A final word

Some companies only occasionally do a license or may be at the beginning of setting up a license project group. In these cases, the company might ask his or her lawyer for advice on how to organize the licensing project group.

The license project group will need the support of a top executive. Without this the license project group will generally not get the resources it needs.

Recognize the different functions to be performed. Then, divide or share these functions depending on available resources, business people, consultants or lawyers:

- Profit Center Group - optimizes the licensing in and licensing out to make money for the company
- Support Group - coordinates technical, legal and accounting aspects of tech transfer
- Administrative Group -handles accounting, record keeping and day to day details of maintaining the license.

V. Conclusion

Technology agreements help people work together while managing the intellectual property of their efforts. These agreements let everyone involved know their rights and responsibilities. Knowing your goal and having the appropriate agreement to help get to the goal and understanding the agreement can avoid nasty surprises. If you know your goals, technology and agreements you will be in a better position to turn that technology into profit.

(20a) Implementation of Process Intensification and Flow Chemistry

Sigurd Buchholz, Bayer Technology Services GmbH, Leverkusen, E 41, 515, Germany

The presentation will highlight selected examples from current industrial development activities at Bayer Technology Services covering innovative chemical route selection, introduction of special technologies, scale-up and small scale manufacturing enabled by process intensification.

Process intensification and flow chemistry are opportunities to open novel pathways in process design and engineering. These step changes can significantly improve efficiency and safety in chemical and pharmaceutical synthesis and reduce capital expenditure. Continuous processes can yield value chemicals at higher efficiency and lower manufacturing costs or even allow the access to new or tricky molecules. A number of highly innovative approaches and protocols are currently being developed in both industry and academia. The presentation will highlight selected examples from current development activities at the Bayer Group. The examples will cover innovative route selection, introduction of special technologies, micro structured equipment design as well as scale-up and modularized approaches. The examples will illustrate the methodology of intensified continuous process and product development demonstrating how novel routes can be indentified and what the benefits and hurdles from a process and manufacturing point of view can be. The presentation will illustrate how scale up effort and time to market can be reduced when moving from "lab chemistry" to pilot and production scale. As reduction of manufacturing costs is becoming increasingly important for the production of even high-value chemicals such as API it will be discussed how routes and technologies can be assessed at a very early stage of development and what the impact of novel processing conditions and continuous manufacturing can be.

(20b) Two Key Challenges for Microchannel Process Technology: Validated Cost of Manufacturing and Manufacturing Infrastructure

Ward E. Tegrotenhuis, Pacific Northwest National Laboratory, Richland, WA

Microchannel process technology (MPT) as a platform for process intensification has been investigated for a broad range of applications and unit operations, including heat exchangers, microreactors, and separations. Several research groups have designed, built, and demonstrated devices utilizing a prototyping process consisting of photochemical machining to pattern metal shims and diffusion brazing to bond stacks of shims to form a device. Alternative processes have been explored that are perceived to be more amenable to high volume production, including stamping, low-temperature brazing, and laser welding. While the performance and utility of MPT devices have been repeatedly demonstrated in the laboratory and technologies are penetrating specific markets, questions remain about the economic viability of MPT for broader markets, particularly those that are more cost sensitive. The two key challenges for answering these questions are validated cost of goods sold for high volume manufacturing and establishing a manufacturing infrastructure. Progress is being made with the first challenge using process-based cost modeling methodologies and validation by equipment suppliers and manufacturers. Alternative scenarios for establishing a manufacturing infrastructure are discussed. A number of case studies for specific devices and applications are described.

(20c) Centrifugal Fields and Process Intensification - the Role of High Gravity Technologies in PI and Microprocessing

Lawrence R. Weatherley, Chemical & Petroleum Engineering, University of Kansas, Lawrence, KS

It is often quoted that a large part of process intensification is about the control of fluid mechanics and perhaps there is no area where this is more the case than in systems which exploit centrifugal fields. The criteria of reduction in residence time, control of mixing and shear, energy efficiency, operating and capital cost, temperature control are all strongly influenced by the fluid mechanics. With this in view, the paper will place the application of centrifugal field based technologies in their historical context with a brief overview. The application of centrifugal fields for the enhancement of separation and contacting processes in multiphase systems has a long history with developments including various designs of rotary contactors developed for enhanced liquid-liquid extraction such as the Podbielnak Centrifuge and the Wesfalia Decantor extractor. Other examples of the application of high gravity fields include the development of cyclone technology for gas cleaning and for solid liquid separations. These are well established and mature technologies with many well proven applications. In the last several years however the concept of process intensification has been strongly driven by a desire to achieve miniaturization in equipment design, which in turn is driven by the resetting of economic, safety, and environmental goals. Another factor is the closer integration of new green chemistry with equipment innovation to achieve viable manufacturing routes for conversion and separations based on new science. The paper will summarize some case studies where high gravity and other centrifugal field based techniques have shown promise in this regard. Secondly some key developments in equipment design with a particular focus on spinning disk technology and related techniques will be reviewed. Finally the importance of simulation techniques in aiding improvements in design and furthering the understanding of the fluid mechanics in rotating equipment will be discussed.

(26a) Simplified Cost Estimation and Its Application to Decision Making: What Do You Really Need to Know?

John Eric Anderson, Core R+D, Engineering and Process Sciences, Dow Chemical, Midland, MI

Simplified Cost Estimation and its Application to Decision Making: What Do You Really Need to Know?

In an environment where new product and process concepts are actively encouraged, competition for both money and people will usually result. While it is true that this is a 'nice problem to have', this forces organizations to make decisions about where to expend limited resources. Valid economic forecasts are required but can be difficult to make in the formative stages of an idea's development since much of the required data is not available. However, in many cases, the impact that a particular unknown has on the economic results can be computed using reasonable ranges of input values and to clarify the economic picture. Case studies will be shared where decision-making was enhanced significantly using economic calculations that were sufficiently simplified that only a few days of effort were required.

(26b) Lessons Learned From SABIC New Processes Commercilization

Fuad Mosa, Sabic Technology Center, SABIC, Riyadh 11551, Saudi Arabia

During the Last 20 years, Sabic has experience within deferent affiliates many new or scaled-up processes where tremendous amount of resources have been devoted to operate and stabilize them. Commercial plant start-up of new process is the real evaluation of how much successful the critical development milestones were accomplished. All lesson learns from the start-up can be linked to the failure of estimating the required activities for each development step. Moreover, As the plant mainly design for steady state with specified margin, initial plant start-up can be considered the most challenging step which most likely not reflected in the design due to the difficulties to set all possible unknown scenario's which can be faced.

In fact, Development of new technology is including many critical milestones which can strongly affect the final plant set-up. Starting from process kinetics and selecting of the important elements for scale-up model, Building the initial concept of the process, choosing the important plant sections which required to be piloting till plant start-up are involving some challenging areas. The final evaluation of the completion for each step can be considered as an important element which can highly affecting the final plant performance.

Typically any new technology runs through an initial learning phases that could last upto many years depending on the complexity of the process. This paper will attempt to share with all sabic and provide the challenges faced in the journey of Many Process commercialization at Sadaf , Ibn Rushed , PK , United, ArRazi and ways to handle these challenges. This will be covered with a clear link between the important process commercialization milestones and start-up challenges faced which are very important for any team working on process development.

Technical Support Benchmarking: A 15 Year Historical Perspective

Christopher F. Morrison, Ph.D.

Senior Research and Development Manager

Nalco, an Ecolab Company

Since 1996, the Nalco Downstream Petrochemicals Research and Development Team has utilized our Technical Support Request (TSR) database to document and resolve more than 4,000 customer petrochemical process issues that could not be resolved at either the local Nalco Sales Engineer or Regional Support level. This TSR database has withstood the test of time, and remains the primary workflow process for the delivery of "high level" support to our customers. Furthermore, this database serves as a rich knowledge repository, as well as a reliable tool to determine market trends and appropriate R&D resource allocation. This paper and associated presentation will discuss how technical support requests have trended with general market conditions, the more recent shift to gas cracking from naphtha cracking, regional shifts in ethylene capacity, unit shifts (e.g. increasing need for support in Primary Fractionators), as well as a statistical analysis of the composition of "typical" ethylene plant deposits.

Technical Support Requests (TSR's) are the primary workflow process for the delivery of "high level" Research and Development (R&D) support to our customers. If a customer problem cannot be resolved at either the Sales Engineer peer level or the regional Industry Technical Consultant (ITC) level, the problem comes into R&D. TSR's are used to solve problems (untreated or competitively held) or to address issues with existing Nalco treated programs. Over the past 16 years, the Nalco Downstream Petrochemicals R&D Team has utilized our TSR database to resolve and document 4,067 such customer petrochemical issues. For 9 of the last 15 years, the annual percent change in TSR's has directionally tracked with the annual percent change in global GDP. For 10 of the last 15 years, the annual percent change in TSR's has directionally tracked with the annual percent change in the total value of chemicals output. By and large, TSR activity is related to general market activity.

The TSR database is crucial to our operations since it's the primary workflow process for Marketing and R&D to deliver sales support. Perhaps more importantly, it is a central knowledge repository and a means for Nalco to share historical site information with our customers that they sometimes lose due to inevitable personnel changes at their customer sites. Furthermore, Nalco has utilized the TSR database to protect our intellectual property, ascertain market trends, and ultimately allocate resources towards issues important to our customers.

While the TSR database started out in the mid-1990's as a North America-centric tool, it was quickly embraced by our global team. As such, it has tracked with global trends in petrochemicals. For example, in 2011 Asia Pacific overtook Europe as the second most demanding region in terms of support behind North America. Recognizing this trend, Nalco redistributed its global Marketing and Sales force to Asia Pacific. Nalco also recognized the need for regional R&D support and constructed state of the art R&D facilities in Pune, India (2009) and Shanghai, China (2008).

Over a 16 year period, there were 1,865 TSR's covering the Olefins industry (excluding Butadiene). During this time frame, the most demanding application was Light Ends (24% of requests), followed by Dilution Steam (19%), Acid Gas Removal (17%), Compressors (16%), and Primary Fractionators (11%). The remaining 13% of requests were mainly around Furnaces, Pygas, and Hydrotreaters. The vast majority (83%) of requests were related to fouling issues, with corrosion (7%), foaming (4%), emulsions/water clarification (3%) and other (3%) rounding out the requests. While Light Ends accounted for the most requests (24%) over the 16 year period, it is worth noting that requests in both Light Ends and Acid Gas Removal have been trending downward, whereas requests for Compressors and Primary Fractionators have been trending upward. Light Ends (Actrene®) and Acid Gas Removal (Arsenal™) are Nalco programs that are particularly well refined and extensively deployed, and thus may account for decreasing support needs.

Since 1996, there have been 203 TSR's for Primary Fractionators. There were "spikes" in Primary Fractionator support in 1999, 2002, 2006, and 2009. With the exception of 1999, each "spike" in Primary Fractionator support was approximately 1 year after a significant increase in crude oil prices: in 2000 crude oil prices spiked 65% and late 2001 saw the beginning of the Afghanistan War; in 2005 crude oil prices spiked 33% (in large part due to Hurricanes Katrina and Rita); and in 2008 crude oil prices spiked 43% - surging to $140/barrel at one point. It is logical to assume that as crude oil prices increase, so too do naphtha prices, thereby compelling mixed crackers to entertain lighter feedstocks. The data we've gathered seems to suggest that there is a "delay" after customers shift to lighter feedstocks before they encounter problems that require high level technical support.

For the 393 TSR's related to Dilution Steam Systems, 37% involved the Dilution Steam Generators, 21% for Quench Water Settler Drums, 12% for Process Water Strippers, 11% for Quench Water Towers, 3% for Feed Saturator issues, 1% for DOX issues and 1% for Blowdown issues. In addition to these, 14% of the TSR's had multiple issues across the Dilution Steam System. Fouling in the Dilution Steam System accounted for the majority of the requests (61%), though corrosion (22%) and emulsions/water clarification (16%) were significant issues as well.

From 1996 through 2011, there were an average of 16.1 TSR's per year related to Compressors – with the notable exception of 2007 when there were 44 TSR's related to Compressors. The vast majority (74%) of these TSR's were for fouling issues. It is interesting to note that for the 290 TSR's related to Compressors, there has been a steady increase in the percentage of TSR's that involve interstage cooler fouling – from roughly 15% in the late

1990's to approximately 40% now. An analysis of which stage of compression shows the most fouling points towards the early to mid-stages of compression, with the later stages of compression having relatively fewer issues.

For the 326 TSR's related to Acid Gas Removal, 82% involved caustic units whereas 18% involved amine units. Despite the fact that amine units represent less than 5% of Acid Gas Removal systems, they appear to require more support as demonstrated by the amount of technical support. 87% of TSR submissions were related to fouling issues, 7% for foaming issues, 3% for corrosion issues and 3% for emulsions. As with Compressors, there was a fairly steady level of requests (20.3 TSR's per year related to Acid Gas Removal) – with the notable exceptions of 2008 and 2011 when there were only 8 and 10 TSR's related to Acid Gas Removal, respectively.

As mentioned previously, the overall trend in Light Ends is one of decreasing support. This may be explained in that Nalco's Light Ends program (Actrene®) is particularly well refined and extensively deployed. However, there were "spikes" in Light Ends support in 2001, 2005 and 2009. It is interesting to note that 2001 and 2009 are years in which the total value of chemicals output suffered a percentage decline (-2.5% in 2001 and -9.8% in 2009) and 2005 was a year that saw many unplanned shutdowns due to hurricanes along the Gulf Coast. During times of decreased throughput, the residence time in Light Ends fractionation increases and this often correlates to increased fouling due to reactive monomers such as butadiene residing in towers for longer periods of time. 41% of TSR submissions were related to depropanizer issues, 25% for debutanizer issues, 19% for deethanizer issues, 9% for depentanizer issues and 5% for condensate stripper issues. Nearly all (95%) TSR's from Light Ends are fouling related, though there are some corrosion issues – namely in the depentanizer – where 15 of the 40 TSR submissions were corrosion related, primarily due to acetic acid corrosion in the depentanizer overhead. Trends in Light Ends are telling: deethanizer and depropanizer support are trending downward, whereas debutanizer support is trending upward.

In conclusion – if the trends hold true – we should expect support requests for both Primary Fractionators and Compressors to rise, with support requests for Light Ends to decline (with the possible exception of debutanizers and during industry downturns). Based on the price of crude oil and naphtha, we can essentially predict when our PrimAct™ program will need more support. This database serves as a rich knowledge repository, as well as a reliable tool to determine market trends and appropriate R&D resource allocation, thereby allowing Nalco to focus on issues of great importance and value to our customers.

Finally, we performed a statistical analysis of 116 representative Light Ends deposits from 2008-2011. The purpose of the statistical analysis was to determine what average levels (and expected ranges) of various elements were. A similar exercise based on 46 representative Acid Gas Removal System deposits from 2008-2011 was also carried out.

Iron (Fe)

80% of light ends deposits contained Iron

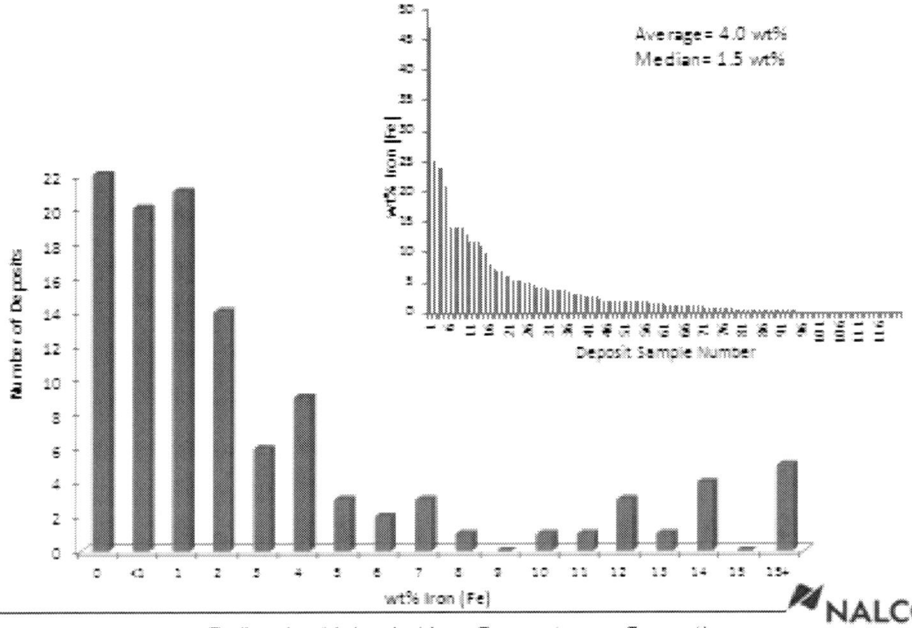

NALCO

Delivering Value in Your Downstream Operation

Nitrogen (N)

23% of light ends deposits contained Nitrogen (>0.3 wt%)

5% of light ends deposits contained Nitrogen > 1 wt%

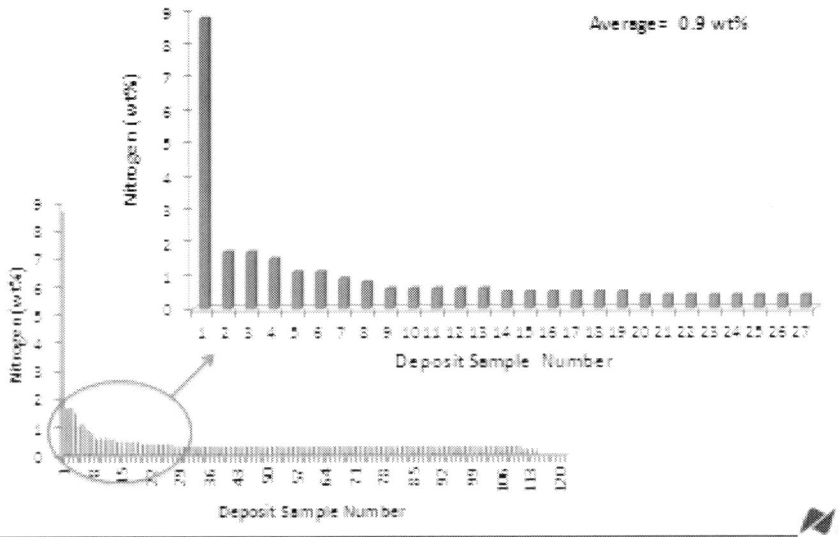

NALCO

Delivering Value in Your Downstream Operation

Sulfur (S)

17% of light ends deposits contained Sulfur

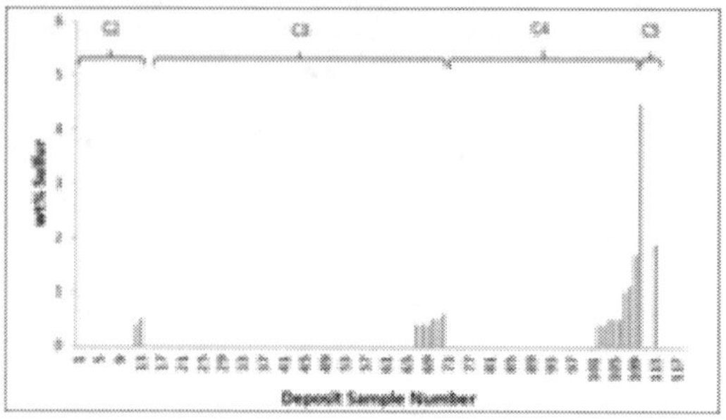

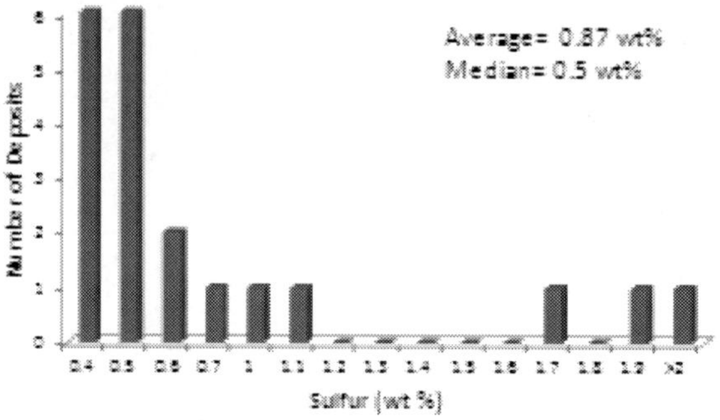

Delivering Value in Your Downstream Operation

NALCO

Silicon (Si)

15% of light ends deposits contained Silicon

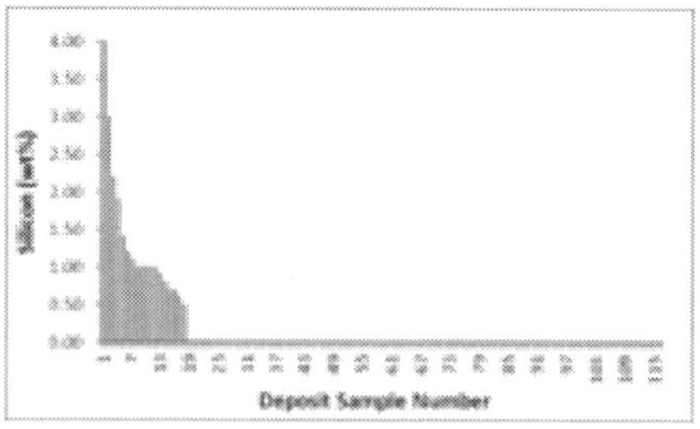

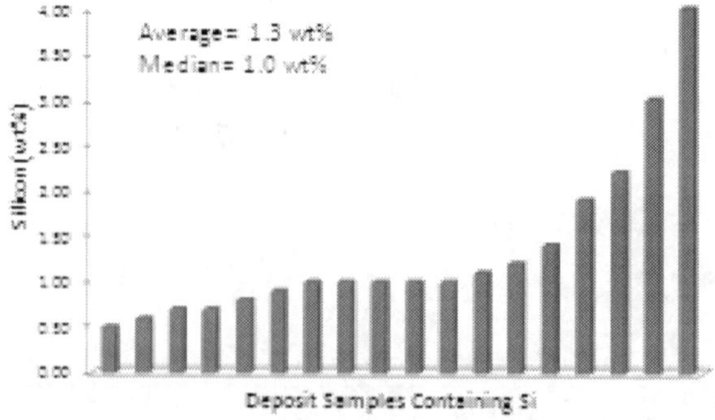

Delivering Value in Your Downstream Operation

NALCO

Additional Elemental Analysis Results

Aluminum

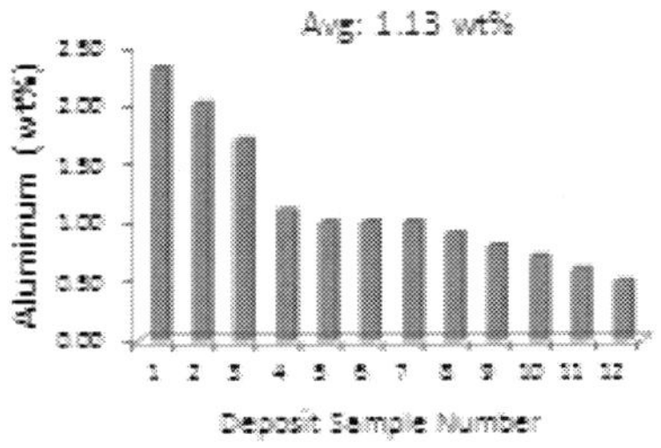

10% of light ends deposits contained Al

Calcium

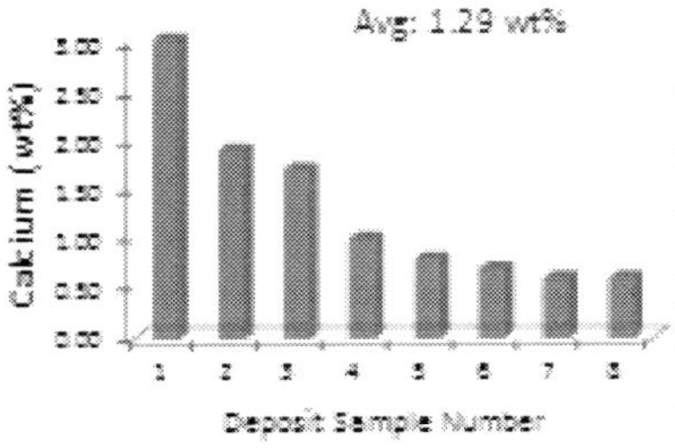

7% of light ends deposits contained Ca

Sodium

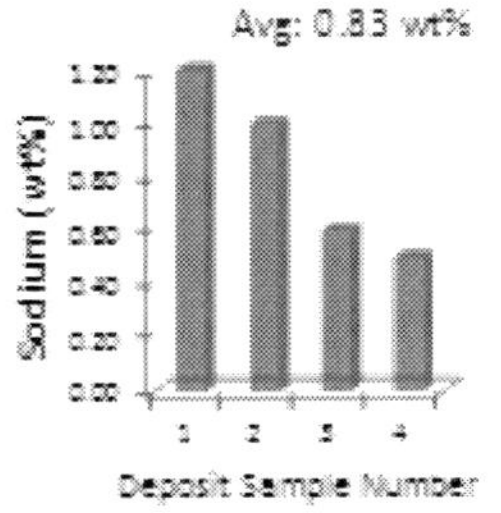

3.5% of light ends deposits contained Na

NALCO

Delivering Value in Your Downstream Operation

Oxygen

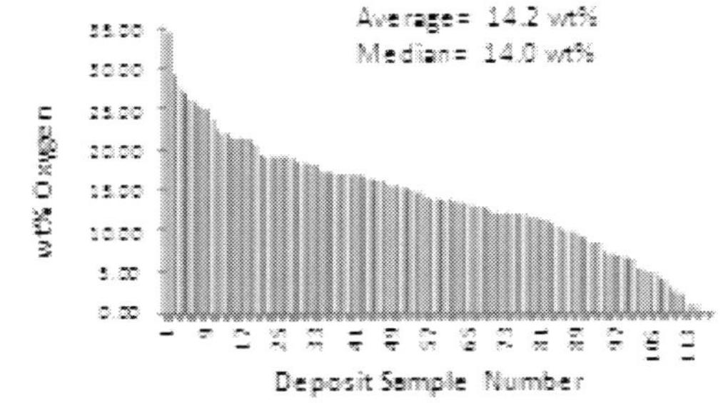

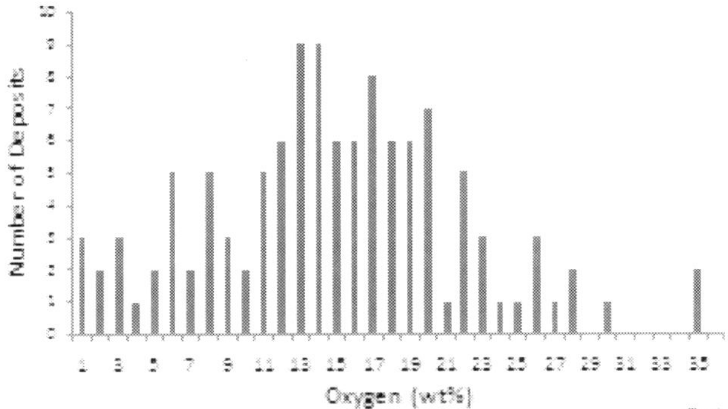

NALCO

Delivering Value in Your Downstream Operation

H/C Ratio

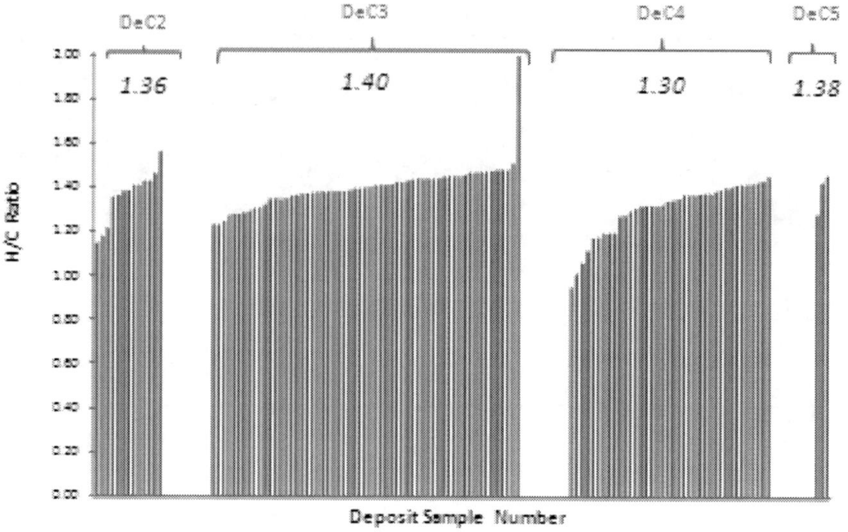

NALCO

Delivering Value in Your Downstream Operation

H/C Ratio

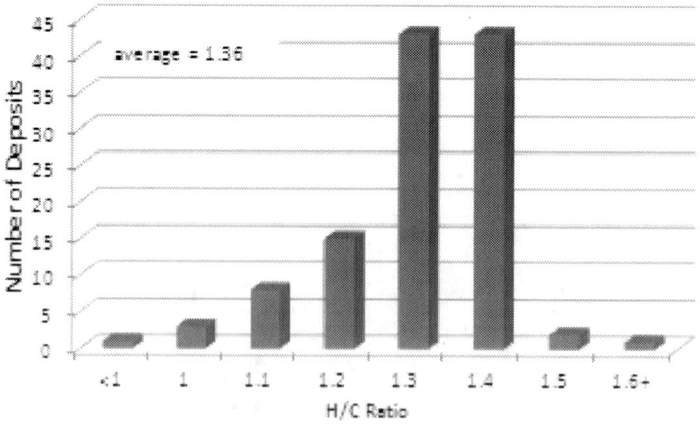

NALCO

Delivering Value in Your Downstream Operation

Loss on Ignition

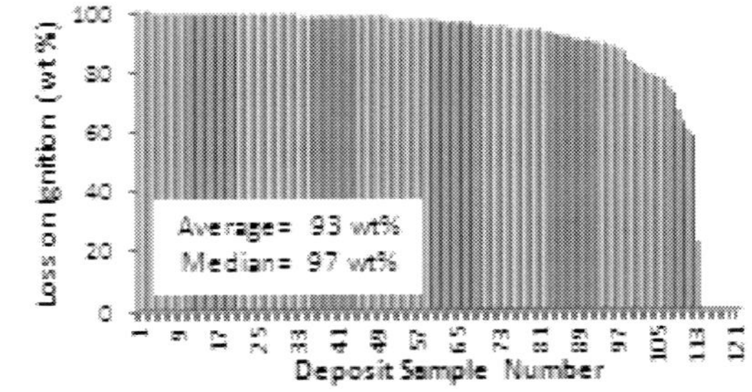

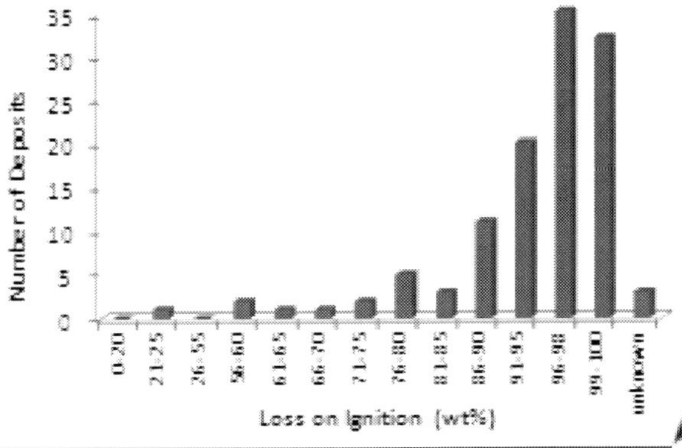

Delivering Value in Your Downstream Operation

Dichloromethane Extractable

Average= 9 wt%
Median= 5 wt%

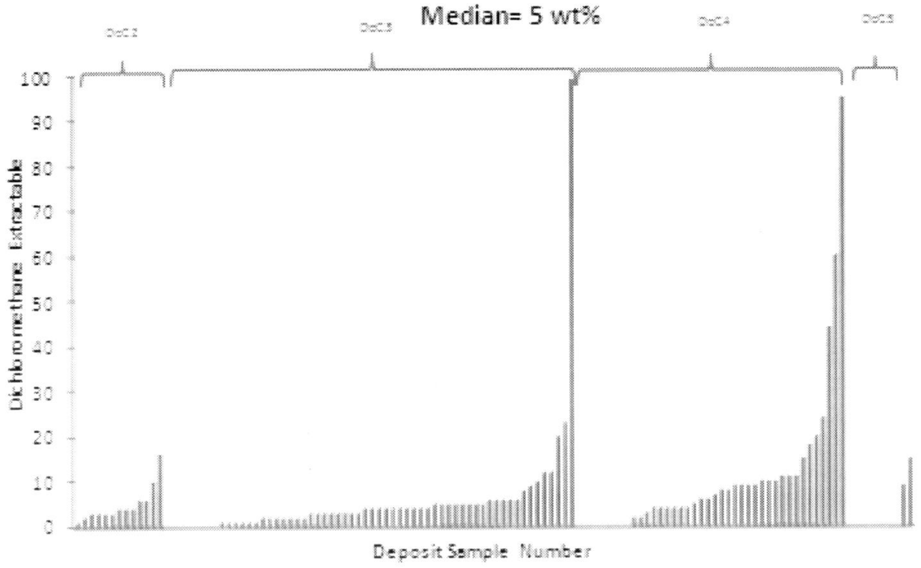

Delivering Value in Your Downstream Operation

Acid Gas Removal Deposits

47 Acid Gas Removal Samples Included in Benchmarking Study (2008-present)

- 89% contained Aldol Polymer
- 92% from Caustic Tower vs. Amine units
- Spent Caustic = 40.4%
- Weak Section = 31.9%
- Intermediate = 8.5%
- Strong = 6.4%
- Other = 12.8%

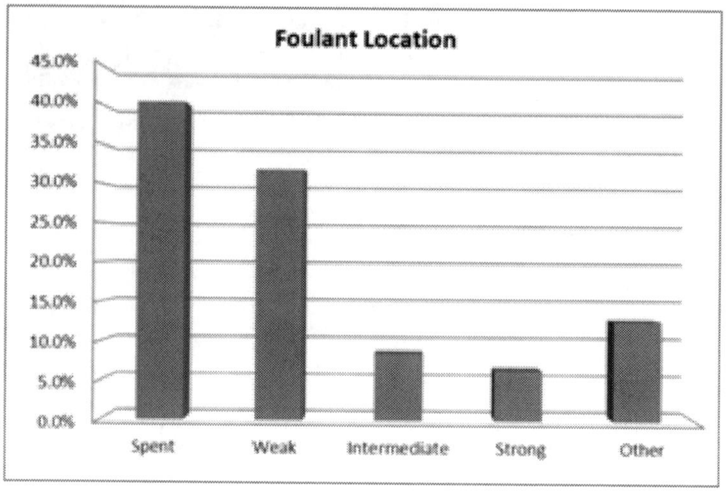

Delivering Value in Your Downstream Operation NALCO

Average Elemental Composition

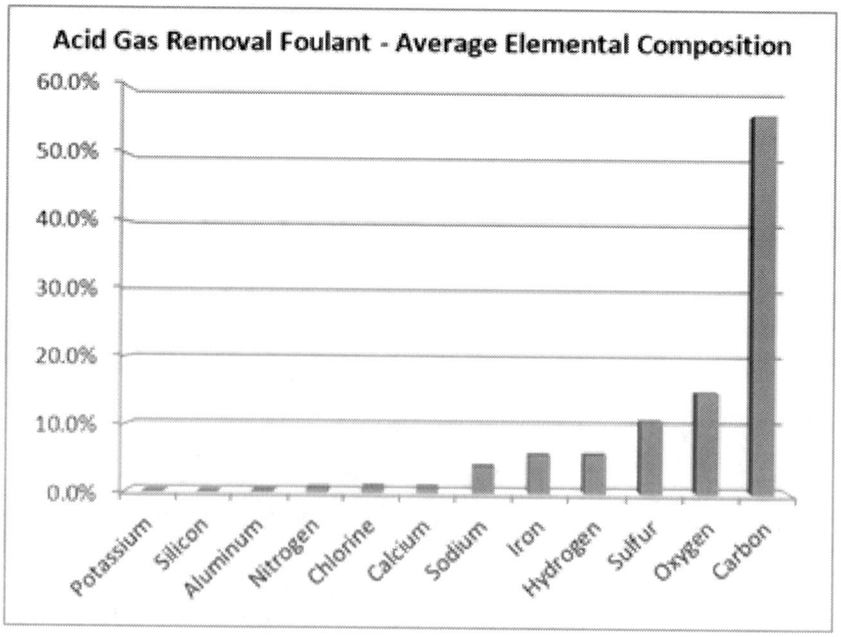

Delivering Value in Your Downstream Operation NALCO

Sulfur and Iron Content

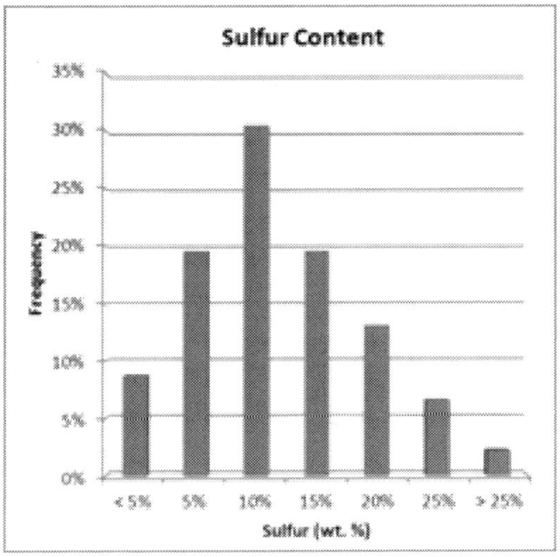

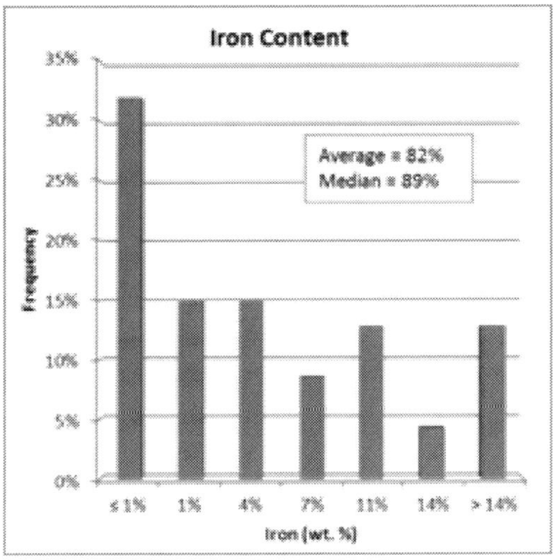

NALCO

Delivering Value in Your Downstream Operation

Oxygen and Nitrogen Content

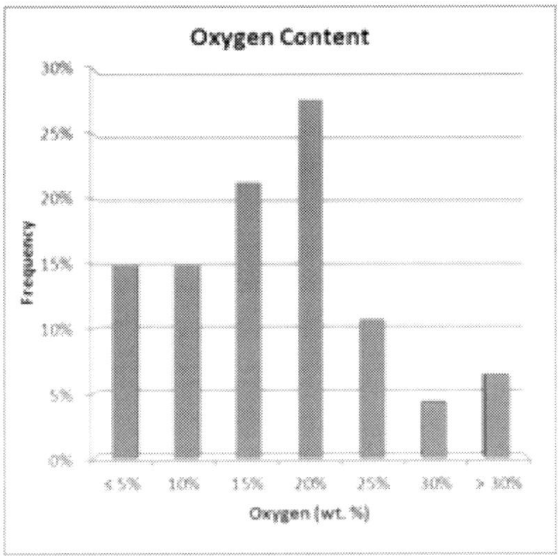

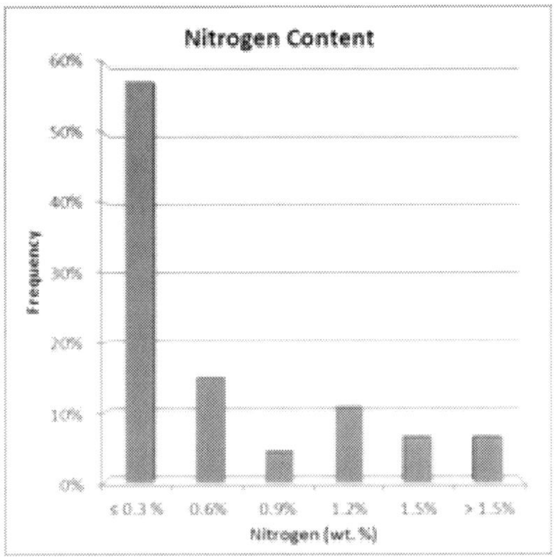

NALCO

Delivering Value in Your Downstream Operation

Solubility and Combustion

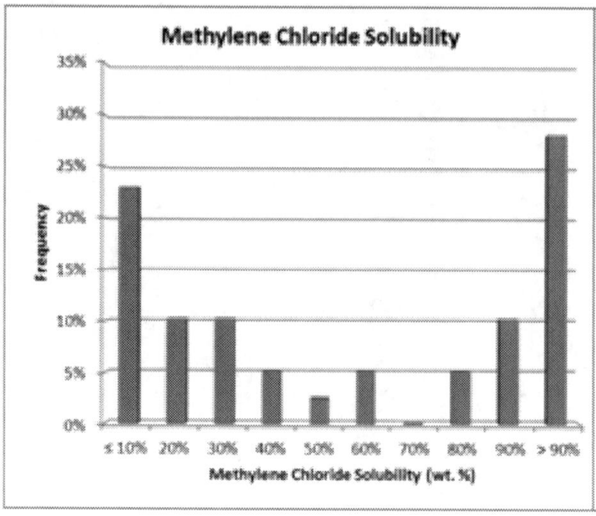

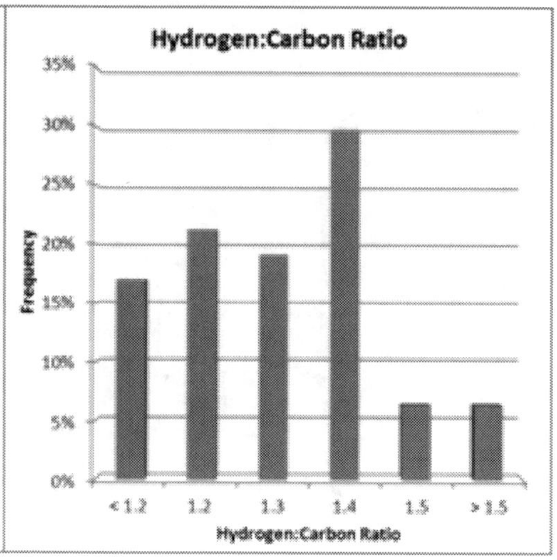

NALCO

Delivering Value in Your Downstream Operation

(26d) An Integrated Approach for Chemical Technology Development

Fuad Mosa, Sabic Technology Center, SABIC, Riyadh 11551, Saudi Arabia

Chemical Technology Development (CTD) projects often span over a decade from initial conception of the idea to commissioning of the commercial plant. To mitigate the high risk involved in developing a new chemical technology from cradle to grave, leading companies have adopted Stage-Gate® System (SGS). However, current as well as future chemical technology development projects are expected to face greater challenges including market uncertainty, less-defined project, better process (product) performance amidst higher technical uncertainties and stricter environmental sustainability requirement. Traditional CTD project management techniques, which are mostly focused on well-defined and predictable projects would not be very efficient in this context. Thus, the need for developing project management techniques that can meet the requirement of next generation CTD projects.

SABIC Technology and Innovation (T& I) Center has recently developed a unique process mapping methodology based on Stage-Gate® System for its next generation CTD projects. The developed process mapping is not a fit-for-all type of Stage-Gate® System used by many companies, but rather customizable to meet the requirement of different technology development projects within chemical and allied industry. The process mapping features innovative concepts, such as rigorous First-Principle Predictive Modeling (FPPM), Design of Experiments (DOE), Quality by Design (QbD), Life Cycle Analysis (LCA) amongst others. The incorporation of these concepts within SABIC SGS is not the only unique aspect, but how these concepts are staged and coupled together to reduce risk in CTD projects while accelerating time-to-market. The newly developed process mapping is currently been used in-house for several of SABIC's CTD projects, and results obtained so far has been astonishing. In this work, some examples of how SABIC's new process mapping could be used to reduced CTD projects' risks while accelerating time-to-market would be demonstrated.

BIOMASS TO OIL: FAST PYROLYSIS AND SUBCRITICAL HYDROTHERMAL LIQUEFACTION

Nicolas Doassans-Carrère, Laboratoire M2P2, Aix-en-Provence, France
Guillain Mauviel, LRGP, Nancy, France
Jean-Henry Ferrasse, Laboratoire M2P2, Aix-en-Provence, France
Jacques Lédé, LRGP, Nancy, France
Olivier Boutin, CEA, Marcoule, France

Abstract

The present paper deals with the comparison of two biomass-to-oil processes: fast pyrolysis and subcritical hydrothermal liquefaction. Using the same biomass (beech sawdust), fast pyrolysis was led thanks to a cyclone reactor (wall temperature between 870 and 1040 K) and subcritical hydrothermal liquefaction thanks to a 150-ml-batch-reactor (temperature between 420 and 600 K). Mass balances and phases analysis (ultimate analysis, higher heating value (HHV), Karl-Fischer, gas chromatographies, H^1 NMR) allow the comparison of both processes and the characterization of pyro-oils (heavy oils, light oils and aerosols), liq-oils (heavy oils and water soluble organics), solids and gas.

For both processes, there is a temperature for which a high conversion of biomass is reached with a minimum of gas mass yield. Up to 62.6 wt% of pyro-oil and 47.0 wt% of liq-oil, on dry basis, were obtained in this work, with oxygen contents and chemical compositions different from one to the other.

Introduction

The global energy demand is continuously increasing, due to the growth of both population and industrialization. The supply of CO_2-neutral energies has to become more important in the future, and bio-fuels from biomass may be a part of those energies.

Different processes have been developed to produce bio-oils from different biomass, among them fast pyrolysis and subcritical hydrothermal liquefaction are studied in this work. Fast pyrolysis can produce up to75%wt of bio-oil (pyro-oil), composed typically of 56%wt of C, 38%wt of O and 6%wt of H [1]. Subcritical hydrothermal liquefaction consists in thermal decomposition of biomass in water in high pressure and temperature condition but below its critical point (647 K and 22.1 MPa). Subcritical water is mainly used today for wet air oxidation [2]. Using subcritical hydrothermal liquefaction, 35%wt of bio-oil (liq-oil) can be reached. This liq-oil, less oxygenated than pyro-oil, is composed typically of 74%wt of C, 17%wt of O and 9%wt of H [1]. Those two types of bio-oils must be upgraded to obtain bio-diesel, thanks to catalytic hydrodeoxygenation (HDO) which reduces the proportion of oxygen.

The aim of this paper is a direct comparison of fast pyrolysis and subcritical hydrothermal liquefaction for the production of bio-oil. This comparison is based on mass balances and products characterization: bio-oils, solids and gas.

Experimental apparatus

In order to compare those processes, experimentations were led with the same biomass, beech sawdust.

Fast pyrolysis

Fast pyrolysis is achieved thanks to the cyclone reactor [3] which is a continuous reactor with an oil recovery system separating three fractions of pyro-oil (Figure 1). Up-stream of the cyclone, the installation is composed of the biomass (tank and screw conveyor and nitrogen (inert carrying gas) feeds, with a mass flowrate ratio between nitrogen and biomass equal to 7.8. Particles, carried by nitrogen, are pyrolysed along the cyclone wall which is heated by induction. A by-pass system allows preheating and equilibrium set-up before biomass feeding. Thanks to the separator properties of the cyclone, solids are isolated from gaseous products and recovered in the solids collector. Gaseous products, evacuating through the upper part of the cyclone, pass through three water cooled heat exchangers in series to recover the first fraction of oil in a collector, "heavy oils" (HO). Then the gas flows through refrigerated coils inside which the "light oils" (LO) fraction is recovered. In order to recover the last fraction, "aerosols" (AE), both electrostatic and membrane filters are used in series. At the exit, clean gases are finally sampled and analyzed by micro-GC.

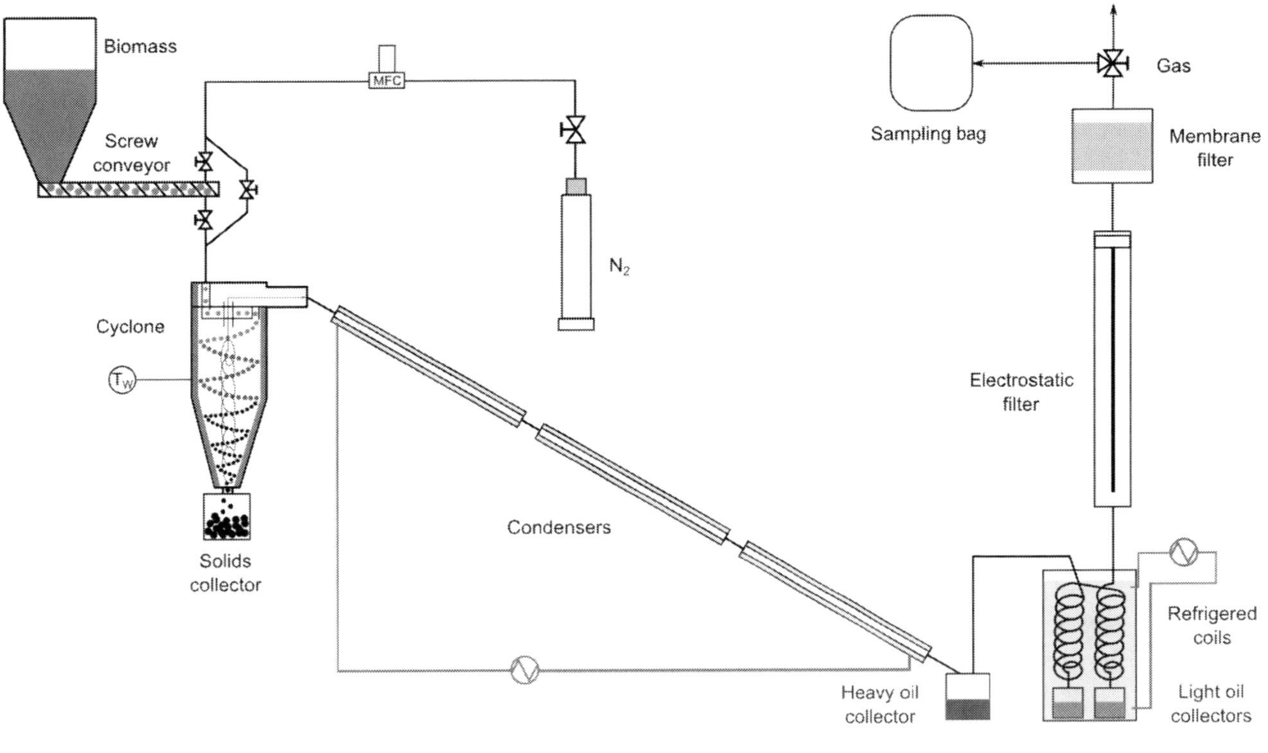

Figure 1. Fast pyrolysis cyclone reactor installation scheme

Subcritical hydrothermal liquefaction

Subcritical hydrothermal liquefaction is achieved in a 150 ml autoclave where biomass (6 g) is introduced prior to the catch of the reactor with the locking straps (Figure 2). Thanks to a volumetric pump, nitrogen or distilled water can be introduced into the autoclave at a desired pressure. After inerting the reactor with nitrogen and setting the initial working pressure (1 MPa), 100 ml of distilled water are added into the reactor, heated by an electric resistance. Once achieved, the working temperature is maintained at the experiment duration time, then the reactor is cooled down by a refrigerating system using air and then water. Once the reactor at the ambient temperature, gases are sampled and analyzed by GC-TCD. The reactor is opened in order to collect the products (liquids and solids). Based on [4], pure acetone is used to rinse the reactor and wash solids. After filtration, three phases are obtained: solids, aqueous phase and organic phase. Solids are placed into a drying oven. Evaporation at reduced pressure is used to remove acetone from the organic phase and water from the aqueous phase, in order to get respectively two oil fractions, "heavy oils" (HO) and "water soluble organics" (WSO).

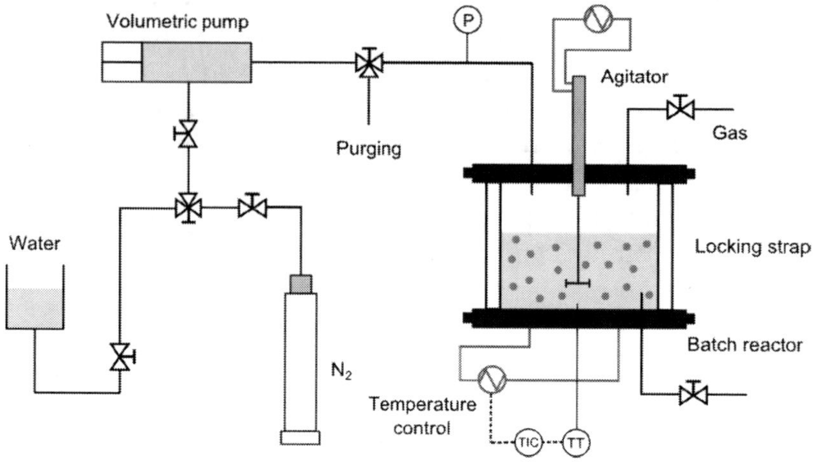

Figure 2. Subcritical hydrothermal liquefaction installation scheme

Results and discussion

Maximum of bio-oil mass yield

For both installations, mass balances are based upon mass measurement of solids and bio-oils, and evaluation of the gas composition. Thanks to the known amount of nitrogen used in those installations (mass flowrate for fast pyrolysis and initial pressure for subcritical hydrothermal liquefaction), the mass of produced gas can be determined from gas concentrations in the gaseous phase. Mass yield are defined from dried biomass, with a biomass moisture of 9 wt.%. It has to be noticed that for fast pyrolysis, after data reconciliation based upon atom balances, mass balances close at 100 % (the initial overall mass balances closures range between 83 and 100 %). No data reconciliation could be made for subcritical hydrothermal liquefaction, and the overall mass balances closures range between 64 and 83 %. This misclosure is often explained by a water production [5].

Figure 3 shows influence of temperature on mass balances for fast pyrolysis and subcritical hydrothermal liquefaction. For both processes, there is a temperature for which a maximum of bio-oil mass yield is reached. For fast pyrolysis, at 970 K, 62.6 wt% of pyro-oils are obtained with 25.7 wt% of gas and 11.7 wt% of solids. For subcritical hydrothermal liquefaction, at 573 K, 47.0 wt% of liq-oils are obtained with 5.5 wt% of gas and 17.8 wt% of solids. This maximum correspond to a high conversion of biomass coupled with a low gas mass yield. The partial conversion of biomass is revealed by ultimate analysis of solids recovered. In fast pyrolysis, the composition of solids recovered at 870 K ($CH_{1.26}O_{0.57}$) is closer to initial biomass one ($CH_{1.48}O_{0.72}$) than chars ones obtained at 1040 K

($CH_{0.24}O_{0.17}$). In subcritical hydrothermal liquefaction, until 528 K, the composition of solids ($CH_{1.50}O_{0.68}$) is similar to the biomass one, whereas from 573 K, solids recovered are closer to chars ($CH_{0.75}O_{0.24}$).

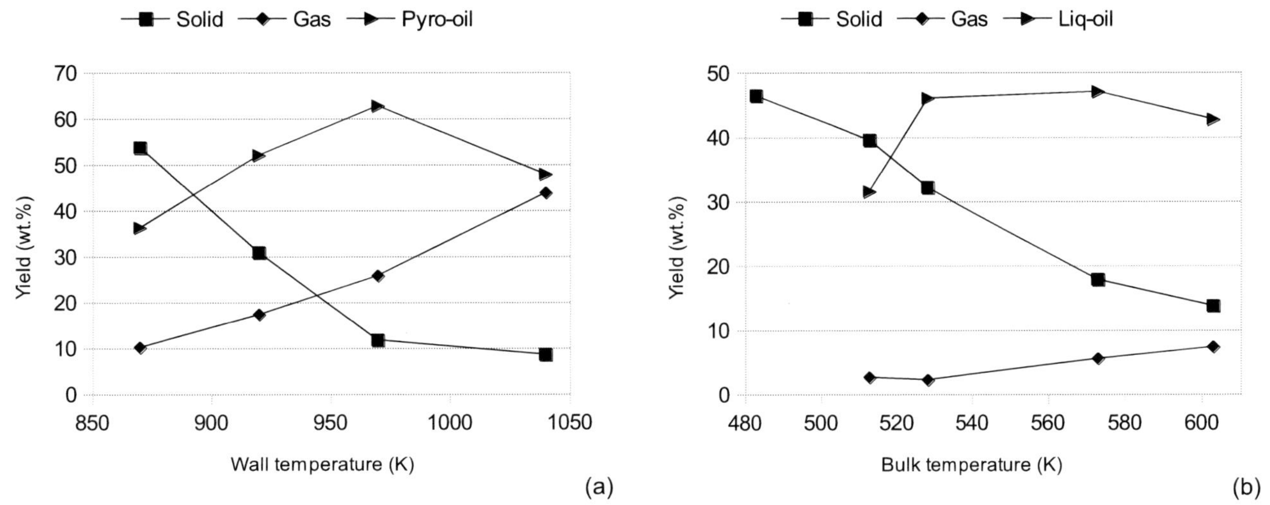

Figure 3. Mass yields evolutions according to temperature for (a) fast pyrolysis and (b) subcritical hydrothermal liquefaction

Products characterization

Apart from ultimate analysis made on solids, which have revealed partial biomass conversion, characterization of gas and bio-oils allows the direct comparison between both fast pyrolysis and subcritical hydrothermal liquefaction.

Gas from fast pyrolysis is a mixture of CO, CO_2, H_2, CH_4, C_2H_4 and others C_2-C_3, with a majority of CO, CO_2 and CH_4 (> 91 wt.%). At the opposite, gas from subcritical hydrothermal liquefaction is mainly composed by CO_2 (> 99 wt.%), with a maximum of 1 wt.% of H_2 at 603 K.

Pyro-oil is the most oxygenated bio-oils with composition of heavy oil, light oil and aerosols of, respectively, $CH_{3.34}O_{1.73}$, $CH_{4.08}O_{1.98}$ and $CH_{1.32}O_{0.46}$. Composition of liq-oil fractions (heavy oil and water soluble organics) are respectively $CH_{1.11}O_{0.43}$ and $CH_{1.65}O_{0.78}$. This difference of oxygen content is partly explained by the water content of pyro-oil. In fact, contrary to liq-oil free of water (consequence of evaporation steps), 40 wt.%, 36.5 wt.% and 6.5 wt.% of, respectively, heavy oil, light oil and aerosols is water.

Gas chromatographies and H^1 NMR confirm this oxygen content difference. Using the same analytical apparatus, a direct comparison is possible between pyro-oil and liq-oil. Those bio-oils have some compounds in common, like phenol, guaiacol or acetic acid, but major compounds are different. For example, aldehydes or carbohydrates like levoglusan, major pyro-oil oxygenated compounds, are absent in liq-oil. At the opposite, hydrogen from alkanes or aromatics, are in higher proportion in liq-oil, indicating longer-chain compounds (and therefore less oxygenated) or higher quantity of compounds from those families.

Finally, ultimate analysis and HHV-measurement, made on solids and fractions of pyro-oils and liq-oils, confirm the use of a correlation Dulong-like between both of those properties [6]. Indeed, liq-oils have higher HHV than pyro-oils, but with a lower mass yield. Energetic balances are close to mass balances with up to 59.5 % and 51.5 % of initial biomass HHV are recovered, respectively, in pyro-oils and liq-oils.

Conclusions

Fast pyrolysis and direct liquefaction of beech sawdust has been studied in the production of bio-oils, through mass balances and products characterization. For both processes, a maximum of bio-oils mass yield is observed which corresponds to a high conversion of biomass coupled with a minimum gas mass yield.

The higher oxygen content of pyro-oils than liq-oils, revealed by ultimate analysis, is explained by the water content and different compositions, shown by GC-MS/FID and H1 NMR analysis.

For future work, those results are used as a basis for energy analysis which must be achieved to ensure the viability of the two processes for the production of bio-diesel, including HDO step. This step is, in particular, simulated thanks to bio-oils analysis like oxygen content and compounds families.

References

1. Huber, G. W., Iborra, S., and Corma, A. (2006). Synthesis of transportation fuels from biomass : Chemistry, catalysts, and engineering. *Chemical Reviews*, 106(9) :4044–4098.
2. Lefèvre, S., Boutin, O., Ferrasse, JH., Malleret, L., Faucherand, R. and Viand, A. (2011). Thermodynamic and kinetic study of phenol degradation by a non-catalytic wet air oxidation process. *Chemosphere*, 84 (9) :1208-15.
3. Lédé, J., Broust, F., Ndiaye, F.-T. and Ferrer, M. (2007). Properties of bio-oils produced by biomass fastpyrolysis in a cyclone reactor. *Fuel, 86*, 1800-1810
4. Sun, P., Heng, M., Sun, S.-H., and Chen, J. (2011). Analysis of liquid and solid products from liquefaction ofpaulownia in hot-compressed water. *Energy Conversion and Management*, 52(2) :924 – 933.
5. Cheng, S. D'cruz, I., Wang, M., Leitch, M., and Xu, CC (2010). Highly efficient liquefaction of woody biomass in hot-compressed alcolhol-water co-solvents. *Energy & Fuels*, 24(9):4659-67
6. Channiwala, S. and Parikh, P. (2002). A unified correlation for estimating hhv of solid, liquid and gaseous fuels. *Fuel*, 81(8):1051

(62b) Pyrolysis Oil and Its Applications

Rajeswar R. Gattupalli, Process Design Development, UOP LLC, A Honeywell Company, Des Plaines, IL, Stanley Frey, Renewable Energy & Chemicals, UOP LLC, Des Plaines, IL and Lance Baird, UOP LLC, A Honeywell Company, Des Plaines, IL

Fuels obtained from biomasses are becoming a valid alternative to the use of fossil fuels, especially in the light of stringent environmental constraints.Fast pyrolysis of lignocellulosic biomass produces a renewable liquid fuel called pyrolysis oil that has a variety of applications as a potential replacement for fossil fuels.One aspect of pyrolysis oil that has been addressed to promote its broader use is improving the stability of pyrolysis oil with respect to viscosity increase over time, making long range transportation and storage possible. Other properties of pyrolysis oil present challenges during its use in standard equipment constructed for petroleum derived fuels. Significant effort has been spent on research and development to improve the stability of pyrolysis oil and for its use in burners and other energy applications.

This presentation is divided into two parts - in the first part, we will give a brief background on pyrolysis oil and its stability concerns, the second part will focus on the rapid thermal processing or RTP process for fast thermal conversion of wood and/or other biomass to high yields of liquid fuels and the various applications of stabilized RTP green fuel.

Catalytic performance of Mn-promoted Ni/MgO catalysts in CO_2 reforming of toluene used as model compound of tar produced from biomass: Influence of preparation methods

Meng Kong, Jin-hua Fei*,Xiao-ming Zheng*

Institute of catalysis, Xixi Campus, Zhejiang University, Hangzhou, P. R. China, 310028

Abstract

Catalytic performance of Mn-promoted Ni/MgO catalysts prepared by different methods was investigated in CO_2 reforming of toluene. By XRD, H_2-TPR and TEM analysis and reforming tests, catalysts made by either conventional impregnation or precipitation methods in which $Mn(NO_3)_2 \cdot 4H_2O$ was used as the start precursor behaved similarly in catalytic performance, Mg_6MnO_8phase was present in all the calcined catalysts which caused worse activities in reforming tests. In contrast, by using a mechanic milling method in which MnO was initially added by physical milling with MgO as a support and then the calcined support was impregnated with nickel nitrate, the main crystalline phase of Mn species was α-Mn_2O_3 which worked as a support, after reduction, Ni can distribute well on both MnO and MgO, in this case, more Ni particles can be obtained because the interaction between Ni and MnO was less strong than that with MgO, therefore, a better performance than Ni/MgO and other catalysts can be yielded.

Keywords: Tar, Reforming, Ni/MgO, Mg_6MnO_8, α-Mn_2O_3

Introduction

Tar is an inevitable by-product during biomass gasification which can pollute environment, block process pipes and decrease calorific value.[1-3]Reforming of tar compounds to syngas is an efficient method to improve the overall energy recovery during gasification while minimizing environmental pollution.[2]

Many researchers have used steam as an oxidizing agent for reforming tar to produce H_2-rich syngas. However, the concentration of steam in the rough production gases from biomass gasification is much less than the concentration of CO_2 which can range from 10 to 30 vol. %.[3,4]Therefore, CO_2 could be a better choice than H_2O for reforming the tar to boost the effective conversion of carbon and minimize the release of CO_2 into the atmosphere by providing more syngas. In our study, toluene was selected as a tar model compound because it represented one of the major tar species **错误!未定义书签。**, after comparing some supported catalysts, Ni/MgO was recognized as an efficient catalyst in CO_2 reforming of toluene, 80% of toluene was transformed to syngas at 600℃with a CO_2/toluene mole ratio of 14.[5]In order to improve the catalytic performance especially at a lower reaction temperature, experimental study and Co-Ni bimetallic catalysts have been investigated.

In this contribution, MnO was used as a promoter, because it showed good capacities in stabilizing Ni particles and eliminating coke in Ni/Al_2O_3catalyst for dry-reforming of methane.[6,7] However, the conventional impregnation and precipitation methods were not suitable in Ni-MnO-MgO catalyst system. Therefore, in this paper, the influence of preparation method of MnO-promoted Ni/MgO catalyst in CO_2 reforming of toluene was investigated, which was helpful for readers to understand the relationship between catalytic performance and catalysts' structure.

2. Experimental Section

2.1. Catalyst preparation

Ni/MgO, Ni/MnO and Ni-MnO-MgO catalysts were prepared by the following methods. In all catalysts the loading of Ni was controlled at 5% and MnO was 10% by weight, in Ni/MnO, the loading of Ni was also 5%, the rest part was MnO.

Ni/MgO was prepared by impregnating with Ni $(NO_3)_2{\cdot}6H_2O$ aqueous solution and MgO overnight, and then dried at 120℃ for 6h, finally calcined at 700℃ for 6h.

Ni-MnO/MgO-CI catalyst was made by a co-impregnation method. MgO was impregnated initially with Ni $(NO_3)_2{\cdot}6H_2O$ and 50 wt.% Mn $(NO_3)_2$ aqueous solution overnight, then followed the above steps.

Ni/MnO-MgO-SI catalyst was made by a sequential impregnation method. MnO/MgO-IM support was prepared first by the impregnation method as same as Ni/MgO, then the modified MnO/MgO support was impregnated with Ni $(NO_3)_2{\cdot}6H_2O$ solution, dried and calcined at the same conditions above.

Ni/MnO-MgO-PI catalyst was made by a precipitation-impregnation method. MnO/MgO-PR support was first prepared by the precipitation method in which50 wt.% Mn $(NO_3)_2$aqueous solution was used as precursor, and Na_2CO_3 was the precipitation agent, and the PH was controlled at 10. After dring and calcination, the modified support was impregnated with nickel nitrate solution as Ni/MnO-MgO-SI did.

Ni/MnO-MgO-MI catalyst was also made by an impregnation method. First, the support of MnO/MgO-MM was prepared by mechanic milling MnO_2 with MgO uniformly, and then calcined at 700℃ for 6h, this support was finially used to prepare Ni-based catalyst by an impregnation method as above.

Ni/MnO_2 catalyst was prepared by co-precipitation with aqueous solution of Ni $(NO_3)_2 \cdot 6H_2O$ and Mn $(NO_3)_2$ and Na_2CO_3 as precipitation agent. Ni/MnO_2 was not made by impregnation, because MnO_2 has a badly worse hydrophilic capacity and it is very hard to squeeze to a suitable size of 0.3-0.45mm..

2.2. Catalyst characterization

The crystalline phases contained in the samples were examined by powder X-ray diffraction system (Rigaku, D/MAX 2500/PC) using Cu Ka radiation (45 kV, 40 mA).

The TEM experiments were performed by using JEM-2010(JEOL Ltd.) TEM instrument operated at 200 kV. An X-ray analyzer for EDS is incorporated into the instrument for elemental analysis under STEM mode. The sample was first dispersed in ethanol and supported on Lacey-formvar carbon on a 200 mesh Cu grid before the TEM images were recorded.

2.3. Catalyst activity test

CO_2 reforming of toluene was carried out at atmospheric pressure in a tubular quartz fluidized-bed reactor having an inside diameter of 9.8 mm. The catalyst (0.5 ml, 0.3-0.45 mm) was first reduced at 700 ^{o}C in H_2 for 60 min and then tested for 450 min at 570 ^{o}C. For testing, liquid toluene was supplied at a rate of 0.3 ml/h byan HPLC pump (ASI, 501) into a constant temperature evaporator (120°C) where a gas mixture composed of 285 ml/min Ar and 14.5 ml/min CO_2 (atmospheric pressure) was introduced at the same time. The flow rates of CO_2 and Ar were measured by mass flowmeters (Brooks, 5850E).

The effluent gases were analyzed with an on-line gas chromatograph (Agilent, GC-6820) equipped with a TDX-101 column and a thermal conductivity detector (TCD) to determine the H_2, CO, and CO_2 concentrations (No CH_4 was detected). Simultaneously, toluene and other aromatic products (just less benzene) were determined with an HP-5 capillary column and a flame ionization detector (FID).

3. Results and discussion

3.1 Characterizations of catalysts

From the XRD patterns of supports used in this study(Fig. 1-A),it was obvious that both Mn/MgO-IM and Mn/MgO-PR supports had distinct diffraction peaks of Mg_6MnO_8 which derived from the rock salt structure of MgO by replacing one-eighth of the Mg^{2+} ions with Mn^{4+} ions and one-eighth with vacancies,[8]where Mn^{4+} ions appeared at the support surface and reacted with MgO to form Mg_6MnO_8,[9] also, due to this formation, MnO could disperse by one single layer on MgO surface.[10] In XRD spectra, there were some diffraction peaks (ca. 42.9°and 62.3°) overlapped together which indicated that both Mg_6MnO_8 and MgO were present on Mn/MgO-IM and Mn/MgO-PR supports. The size of Mg_6MnO_8 particles in either Mn/MgO-IM or Mn/MgO-PI was similar at 14-16 nm (Mg_6MnO_8 (311), $2\theta = 35.4°$). However, for MnO/MgO-MM support, besides the characterized peaks of MgO, the dominant phase of Mn species was α-Mn_2O_3 with a particle size of 32 nm (Mn_2O_3 (211), $2\theta = 32.9°$) which came from calcination of MnO_2.[11]

In XRD patterns (Fig. 1-B), a physical mixture composed of 50wt.%NiO and 50wt.%MgO without calcination exhibited evident difference of NiO and MgO signals in the range of 70-80°, so this region can be used to determine NiO and MgO.[12] However, in Ni/MgO catalyst, no NiO phase was detected, and the diffraction peaks were close to pure MgO, which indicated NiO dispersed well on the support of MgO owning to solid solution structure between NiO and MgO.[12] In Ni/MnO catalyst prepared by co-precipitation, there were also no NiO peaks besides α-Mn_2O_3 with a particle size of 58nm (Mn_2O_3 (211), $2\theta = 33.1°$), which suggested that Ni species also had a good dispersion.

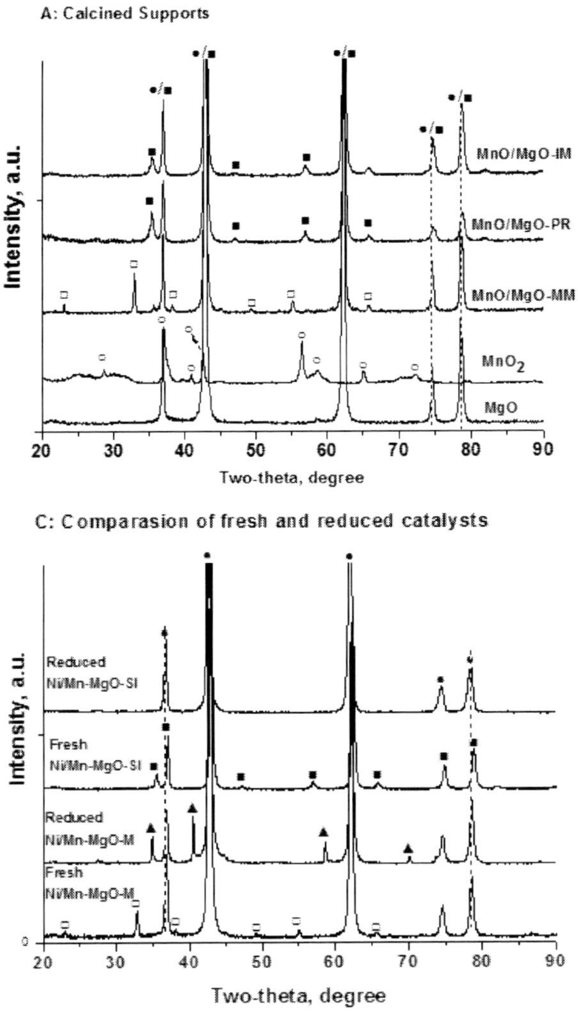

Fig. 1 XRD patterns of (A) Calcined supports, (B) Calcined catalysts, (C) Reduced catalysts. Symbols: □Mn_2O_3, ○MnO_2, ▲MnO, ⊕NiO, ●MgO/or Ni-Mg-O, ■Mg_6MnO_8.

For the catalysts made by $Mn(NO_3)_2$ as the start precursor (SI, CI, PI), the crystalline phases of Mn and Mg species in XRD patterns did not change obviously in which $Mg_6MnO_8$and MgO were the main materials. But because the catalysts were calcined at 700℃ for 6h again, the particle size of Mg_6MnO_8 was increased slightly from 14-15nm to 18-20nm. Of particular interest in fig.1-B was that there were no NiO peaks present on catalysts, which might because NiO was covered byMg_6MnO_8 phase. As for Ni/MnO-MgO-MI catalyst, α-$Mn_2O_3$was also the main Mn species with a particle size of 36nm, no NiO and other manganese-contained phases.

In sum, Mg_6MnO_8 presented on all the catalysts with $Mn(NO_3)_2$ as the start precursor, whereas, Mn_2O_3 was the main phase of manganese on Ni/MnO-MgO-MI catalyst. So Ni/MnO-MgO-SI can be selected as a representativecatalyst to compare with NiO/Mn-MgO-MI.

When these two fresh samples were reduced at 700°C for 1h in H_2, the main states of Mn changed (Fig. 1-C). It was clear that Mg_6MnO_8 disappeared in the reduced Ni/MnO-MgO-SI catalyst without other Mn species, which suggested that the reduced Mn species would have a good dispersion on MgO surface. However, Mn_2O_3 was reduced to MnO on Ni/MnO-MgO-MI after reduction.

TEM pictures gave very different morphology of the refuced catalysts in Fig. 2. Fig. 2-A was the reduced Ni/MnO-MgO-SI, it was clear there were some squares on the MgO surface which was not showed in Fig. 2-B of the reduced Ni/MnO-MgO-MI. The different images derived from the different Mn species definitely. Noted that, in the EDS analysis of reduced Ni/MnO-MgO-MI, the Mn element seemed absent, that was because MnO has a high density which easily deposited in ethanol before TEM characterization.

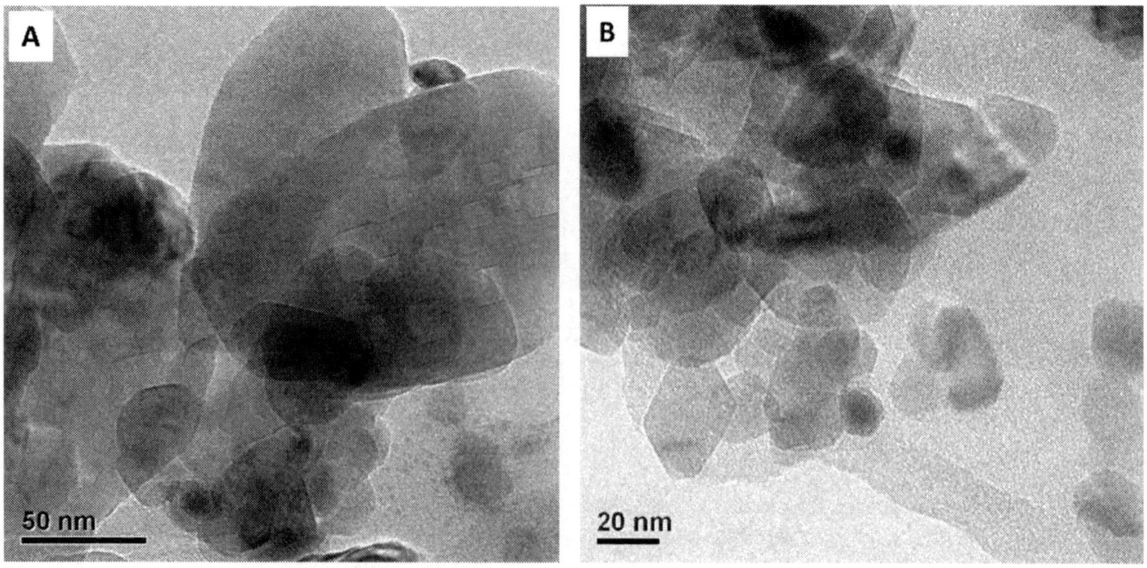

Fig. 2 TEM pictures of(A) reduced Ni/MnO-MgO-SI (B) reduced Ni/MnO-MgO-MI.

3.2 Activity of the employed catalysts in CO_2 reforming of toluene

The performances of catalysts mentioned above were drawn in Fig. 3. As (is) seen from these curves, over Ni/MgO catalyst without any promotes, the initial conversion of toluene can reach to 72.5% (data at 30min), comparably, it will be just 37% when just MnO severed as the support. However, for the Ni-MnO-MgO catalysts with $Mn(NO_3)_2$ as the precursor, the conversions of toluene were no more than 20%, among them, Ni/MnO-MgO-PI had a relatively higher activity with 22.5% of toluene converted at initial time, but it decreased quickly to less than 10% after 5h test. Interestingly, by using MnO_2 as the precursor, a highest and stable performance with 81% of toluene converted can be obtained over Ni/Mn-MgO-MI catalyst. For Ni-MnO-MgO with $Mn(NO_3)_2$ as the precursor, the total conversion of toluene were no more than 10%, including （that） 1.5%-2.5% of toluene was converted into benzene, they had a poor capacity for deep cracking of benzene. However, the proportion for toluene to benzene was also 0.8 % over Ni/MgO, 0.4% over Ni/Mn-MgO-MI, it can be considered that the existence of Mn promotes the conversion of toluene.

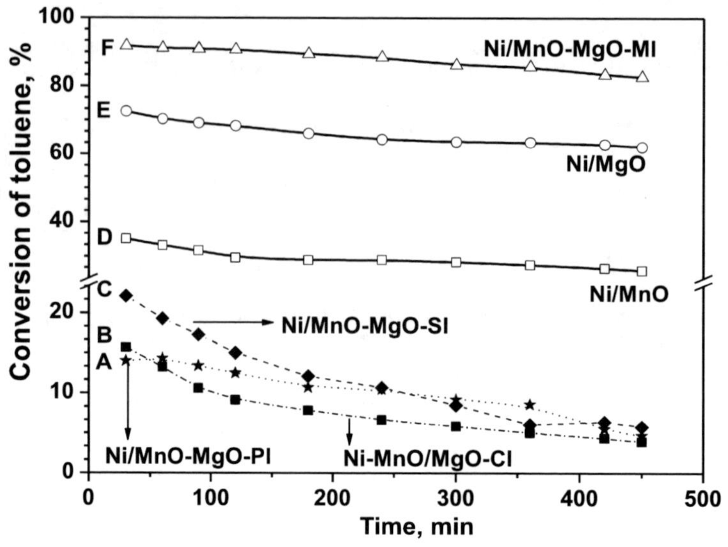

Fig. 3. The conversion of toluene over the catalysts on stream. A, Ni-Mn/MgO-PI; B, Ni/Mn-MgO-CI; C, Ni/Mn-MgO-SI; D, Ni/MnO; E, Ni/MgO; F, Ni/Mn-MgO-MI. Reforming conditions: 0.5ml catalyst, 0.005ml/min toluene, 14.5ml/min CO_2, 285ml/min Ar, GHSV= 36000 h^{-1}, temperature was 570℃.

Conclusion

Mn-modified Ni/MgO was investigated for CO_2 reforming of toluene. It was found that the preparation method played a crucial role in catalytic performance. Using $Mn(NO_3)_2$ as the precursor, Mg_6MnO_8 formed during calcination, which made the catalysts obtain a worse activity, however, using MnO_2 as precursor, the interaction between Mn species and MgO became weak, MnO can work as a support to distribute Ni particles, which benefited the reforming reaction.

Acknowledgements

This research work was supported by the Hi-Tech Research and Development Program of China (2007 AA 05Z415) and the Natural Science Foundation of Zhejiang Province (Y407113).

Reference

[1] Coll, R., Salvado, J., Farriol, X., Montane, D., Fuel Process. Technol., 2001, 74, 19-31.

[2] Zhang, R.Q., Brown, R.C., Suby, A., Cummer, K., Energy Conversion and Management. 2004, 45, 995-1014.

[3] Garcı´a. L., Salvador, M.L., Arauzo, J., Bilbao, R., Energy Fuels., 1999, 13, 851-859

[4] Li, X.T., Grace, J.R., Lim, C.J., Watkinson, A.P., Chen, H.P., Kim, J.R., Biomass Bioenergy. 2004, 26,171-193

[5] Kong, M., Fei,J.H., Wang, S., Lu, W., Zheng, X.M., Bioresour. Technol. 2011, 102, 2004-2008.

[6] Nandini, A., Pant, K.K., Dhingra, S.C., Appl. Catal. A: Gen. 2005; 290: 166-174

[7] Seok, S.H., Choi, S.H., Park, E.D., Han, S.H., Lee, J.S., J. Catal. 2002, 209, 6–15.

[8] Taguchi, H., Nagao, M., J. Mater. Sci. Lett. 1991,10, 658 - 659

[9] Guilliat, I.F., Brett, N.H., J. Chem. Soc. Faraday Trans. 1972, 68, 429.

[10] Yao, J.S., Meng, M., Luo, J.Y., Zha, Y.Q., Xie, Y.N., Hu, T.D., Liu, T., J. Mol. Catal. (China), 2006, 20, 300-305

[11] Ardizzone, S., Bianchi, C.L., Tirelli. D., J. Electroanal. Chem. 1997, 425, 19-23.

[12] Ruckenstein, E., Hu, Y.H., Appl. Catal. A: Gen. 1999,183, 85-92.

(62d) Direct Inter-Processes Integration for Energy Efficiency Optimization and GHG Emissions Reduction in Oil Refining

Mahmoud Bahy Noureldin, Process and Control Systems, Saudi Aramco, Dhahran, 31311, Saudi Arabia and Faisal AlMousa, Process and Control Systems Department, Saudi Aramco, Dhahran, Saudi Arabia

Industrial communities are invited to conserve energy resources not only to give a chance to pre-developing countries and future generations to grow and prosper but also to preserve the environment. Energy efficiency optimization is a quick answer to energy-based GHG emissions reduction. It enables us get two targets with one arrow. From GHG emissions outlook, energy efficiency optimization applications in industry can help us get out tens of thousands of cars abstractly from our roads and practically from our air every year.

Since the emanation of the pinch technology and its evolution to pinch analysis technique and the adaptation of MINLP techniques for process synthesis of mega complexes, the direct inter-processes integration has been considered impractical. For inter-processes "energy" integration many arguments against the direct inter-processes integration are certainly valid.

Plants are normally have different start up and shut down times; the processes can work at partial loads; the processes can have seasonal changes in its operating conditions; utility systems, heaters, coolers and HEN capital will not be reduced and may be even increased due to differences in processes/plants schedule and operation philosophy; the disturbance in one process plant can propagate to another one if they are mass and/or energy integrated; the distance-time/velocity lags in process will affect the controllability of integrated processes; the geographical distances among processes will cost us energy in pumping or compression and capital in piping; plants' safety might be impacted due to the travel of a fluid from one "hazardous" area classification to another; the fear of leakage and so on. Therefore, direct inter-processes integration while is beneficial to energy conservation and the GHG emissions reduction as will be demonstrated in the paper, is still to date almost ignored in the industrial community. Due to those concerns most of the current methods for inter-process integration are indirect and conducted using buffer systems. Buffer systems are either steam system, boiler feed water, hot water system, hot oil system or sometimes combinations of them. The industrial community does agree that, direct integration approach in inter-process integration (between several plants) is efficient and can render more saving in energy consumption and energy-based greenhouse gas emissions beyond the maximum possible using intra-integration only. In this paper, we demonstrated that huge potential for energy consumption and GHG emissions reduction in oil refining can be attained through smart integration among processes. The paper has two folds of results. First, while we agree with the validity of some of the arguments against the direct inter-processes/plants integration we believe and will demonstrate that we still have room for improvement in energy efficiency enhancement and energy-based GHG emissions reduction. We will identify at the targeting phase best possible scenarios for inter-processes integration and the most cost effective solution(s) now and in the future via "Plants' Smart Matching", without the discouraging solutions which require many inter-connections among plants as sometime reported in literature. Second, the intuition that mandates sophisticated inter-plants integration to achieve good benefits, the more you integrate among the processes (new mega sites consists of integrated refining, chemical, petrochemical and plastics plants), the better you save in energy consumption is not always true. The paper presents those two results and demonstrates the benefit that could be obtained from the direct inter-processes integration through two industrial case studies.

(62e) Energy Efficient Control Strategy to Operate Post- Combustion CO2 Capture in Response to Power Plant Load Reduction

Sepideh Ziaii, Gary T. Rochelle and Thomas F. Edgar, Department of Chemical Engineering, The University of Texas at Austin, Austin, TX

Energy Efficient Control Strategy to Operate Post- Combustion CO_2Capture in Response to Power Plant Load Reduction

Increasing concern about global warming has motivated the development CO_2capture and sequestration as a potential option for reducing emissions from PC and IGCC plants. One of the mature technologies that can capture CO_2emissions from a coal-fired power plant is amine-based absorption/stripping, which has the advantage of being added to an existing power plant. However, this technology is a very energy intensive, which can reduce the efficiency of power plants by 30%.

In thistechnology the aqueous amine solutioncountercurrently$_2$from the flue gas by exothermic chemical reactions and significantly reduces the CO_2concentration in the gas stream exiting the absorber. The CO_2-rich amine solution from bottom of the absorber is sent through a cross exchanger and directed to the stripper. The stripper temperatureis maintained$_2$reaction and liberate CO_2. The lean solution leaving the bottom of the stripper iscooled and $_2$. The concentrated CO_2stream from the top of stripperis compressed

Due to high-energy demand that is mostly attributed to solvent regeneration and CO_2compression, many research efforts seek to find the most energy efficient steady sate design . However there are few studies havebeen performed on minimizing energy consumption during possible operational scenarios for an existing capture plant. This study has simulated

This work incorporated models of MEA absorption/stripping process, CO2flowsheet in the framework of Aspen Custom Modeler. Absorber and stripper are packed columns modeled as multi-segment components with time variant mass and energy balance equations. Each segmentis modeledwas modeledare modeled

During boiler load variation in power plants, so many process variables are subject to change. One of the important variables influencing downstream capture plant operation is the property of flue gas entering the absorber. Based on discussion in literature on reduced load operation of coal-fired power plants, itcould be assumed that as load reduces, the rate of flue gas is reduced accordingly while the composition and temperature are constant. The other issue resulting from power plant load change is related to thesteam rate1st

Regarding energy efficient operation, the simulation results demonstrated that a variable speed compressor is much more beneficial than a constant speed one. Adjusting compressor speed reduces lost work by pushing the compressor away from surge condition and avoids additional energy use due to recycling gas in the compressor. In addition, as total steam rateis reduced in power cycle, the reboiler steam pressure and flow rate is reduced and consequently CO22heat duty equivalent work; however, the temperature should not exceed 1202is limited by maximum compressor speed, which is typically 120% of rated speed.

Another potential processvariable which can be manipulated to minimize reboiler lost work and save the compressor from surging is solvent circulation rate. At fixed reboiler temperature, there is an optimum solvent rate that minimizes lost work and maximizes CO2is set by compressor and pump operability. The minimum flowis always setis operated

Based on minimization of total lost work by compressor speed and solvent circulation rate performed in the presence of operational constraints associated with pumps, compressor and solvent thermal degradation, this study gives a practical optimum CO_2removal over 40%-100% operating load range. For this range of load change in an MEA plant designed for 90% removal, the plant can remove up to 94% of inlet CO_2. The practical minimum removal over that load range is about 85%, where the reboiler temperature reaches its maximum allowable value and the lost work is at a minimum.

Reducing the load of power plant that reduces extracted steam rate in the capture results pushes the compressor operating curve to the surge limit such that at minimum load (40%) , adjusting solvent rate and compressor speed no longer moves the operating curve away from

Based on energy minimization in response to power plant load reduction, two control strategies were found to operate the capture close to minimum energy path at each desiredremoval .

1. Ratio control between reboiler steam rate and CO_2rate in rich solution or
2. Controlling lean loading to vary proportional to the CO_2removal change.

In the future we plan to observe and compare dynamic performance of these two control strategies.

(62f) Study on Cascade Refrigerant and Mixed-Refrigerant Systems for Ethylene Plants

Ha Dinh, Jian Zhang and Qiang Xu, Dan F. Smith Department of Chemical Engineering, Lamar University, Beaumont, TX

Refrigeration system holds an important role in most chemical/petrochemical processes. A refrigerant system transfers heat from low-temperature sources to high-temperature sinks through vapor-compression cycles at the expense of mechanical work, magnetism, laser or other means. In many industrial processes, the refrigeration systems are employed for chilling and freezing the final liquid/solid products or liquefying the gas intermediates/products for the downstream cryogenic separation. Optimal synthesis of these systems is attractive not only because of their economic benefits but also because they incorporate most of energy transfer operations in widespread use in the chemical industry. The traditional cascade refrigeration system (CRS) used in ethylene plants includes multiple refrigerants and multiple temperature and pressure levels. Each component subsystem includes a compression section, a condensing section and several temperature levels with sub-coolers, expansion valves, evaporators and flash drums. Recently, mixed-refrigerant system (MRS) has shown some advantages comparing to the single-component one. A MRS uses a mixture as the refrigerant instead of several pure components as in conventional cascade refrigeration systems. This system operates within smaller temperature differences at the lower limit; therefore, it leads to a smaller increase in entropy and consequently a smaller loss in energy. Besides, an MRS also requires a simpler unit configuration and fewer maintenance problems.

In this study, both types of refrigeration systems used in ethylene plants are studied based on exergy analysis, a powerful method to explore system energy loss below ambient temperature. The procedure involves three consecutive steps: Thermodynamic Analysis, Synthesis Model Development and Solving, and Solution Validation and Evaluation. Rigorous steady-state models of both refrigeration systems are firstly constructed and employed to generate Exergy-Temperature (B-T) chart in order to analyze their thermodynamics and energy distributions. An exergy-based mixed-integer linear programming (MINLP) model is developed to maximize the energy efficiency of each system, where multiple refrigerants with multiple recycling loops are simultaneously addressed and all of cooling demands are fulfilled. The optimal solution is subsequently examined by conducting rigorous simulation to check their feasibility and consistency. Economic evaluation is also carried out from the calculated capital cost of new equipments and annual operating cost. The efficacy of the developed methodology is demonstrated by two case studies: a cascade refrigeration system and a mixed-refrigerant system used in two ethylene plants.

COMBINATION FILTRATION
OF PRESSURE, VACUUM AND CLARIFICATION TECHNOLOGIES FOR OPTIMUM PROCESS SOLUTIONS

Barry A. Perlmutter, President & Managing Director
BHS-Filtration Inc.
9123-115 Monroe Road
Charlotte, North Carolina 28270
Phone: 704.845.1190
Fax: 704.845.1902
E-mail: barry.perlmutter@bhs-filtration.com

INTRODUCTION

Filtration experts, over the years, have discussed combination filtration and debated its definition. In the realm of cartridge filtration, simply defined, a combination filter is one that does at least one other processing job at the same time as filtering a suspension. For example, this could be carbon canister which removes both suspended and dissolved components. In water applications, a combination filter removes bacteria, sediment, chlorine taste and odor, and scale. In lubrication oil filtration, combination filtration refers to full-flow and by-pass flow filtration. Finally, for small scale process filtration, combination filtration is installing bag and cartridge filtration systems in series.

There is, however, a new definition of combination filtration that transcends the standard approach and will assist process engineers with trouble shooting and "idea-generation." The approach relies upon the slurry analysis and testing to uncover the "process symptom" and then develop a process solution called "combination mechanical slurry conditioning and filtration." This article discusses slurry testing and process analysis and then illustrates seven (7) case history example installations showing "combination mechanical slurry conditioning and filtration."

TESTING TO DETERMINE "COMBINATION MECHANICAL SLURRY CONDITIONING AND FILTRATION"

Overview of Bench Top Testing in the BHS Laboratory

The BHS bench top testing is conducted using the BHS Pocket Leaf Filter, as shown in Figure 1. The testing will analyze cake depths, operating pressures, filter media, washing and drying efficiencies and qualitative cake discharge. The data collection sheets are shown in Figure 2.

Typical Example of "Combination Mechanical Slurry Conditioning and Filtration" from Bench Top Testing in the BHS Laboratory

The typical testing program is shown in Figure 3. The results and summary of the testing illustrated that the most effective approach for the process filtration is a two stage process.

The primary filtration is by candle filters on a continuous basis. These filters are able to retain particles down to 0.5 microns and discharge these solids as a concentrated sludge/slurry. In order to reduce utilities, no drying or washing is necessary in the larger candle filters.

The concentrated slurry will then be filtered on a secondary pressure plate filter to recover the remaining solvent and discharge the dry solids. The secondary filtration is a batch process which will incorporate filtration, counter current washing, drying, and dry cake discharge. The overall result is a very dependable process with high quality filtrate and minimal utility air and water usage. Each of these processes for the primary and secondary filtration has been tested to confirm that the overall design requirements can be met.

COMBINATION FILTRATION TECHNOLOGIES OF PRESSURE, VACUUM AND CLARIFICATION

There are, without doubt, many technologies that can be applied in combination, including the use of chemicals such as flocculants and coagulants, and which would be the subject of a filtration-chemical handbook/textbook. These already exist in the marketplace. However, from a practical viewpoint, for the basis of this article, a brief description of five (5) types of filtration technologies is presented below based upon general operating conditions at chemical plants.

High-Solids Slurries: Continuous Pressure Operation

The Rotary Pressure Filter technology provides for thin-cake, continuous production in a single unit. Filtration is conducted via pressure of up to 90 psig. Positive displacement washing or counter-current washing follows filtration. Of course, multiple washing steps as well as solvent exchanges, steaming and extraction can also be accomplished. Finally, the cake is dried by blowing hot or ambient-temperature gas through the cake. The Filter has a uniquely designed discharge system, which provides for atmospheric discharge from pressure filtration. After automatic cake discharge, the filter cloth is washed; the clean filter cloth the re-enters the feeding / filtration zone thereby continuing the process. All solvent and gas streams can be recovered separately and reused in the process to minimize their consumption.

As for the operation, each process zone (typically 5 - 7 zones) is isolated by a separating element. The pressure in the specific zone can be adjusted to meet the process parameters. The separating elements are sealed to the rotating drum via a gas membrane pressure seal. This drum is sealed to the outer housing by drum packing. The drive system with self-sealed roller bearings provides for high stability and low stress on the drive.

High-Solids Slurries: Continuous Vacuum Operation

High-solids slurries can be defined as up to 50 – 55% solids in the slurry feed. In some cases, for high solids applications, the slurries can be better handled using vacuum filtration rather than pressure filtration. An example of a continuous, thin-cake technology is a Continuous – Indexing Vacuum Belt Filter. This technology consists of fixed vacuum trays, continuously feeding slurry system and indexing or step-wise movement of the filter media. The filter media is indexed by pneumatic cylinders located on the exterior of the unit. The pneumatic operation and fixed trays eliminates a motor and variable speed drive, there are no rails/rollers, and no rubber carrier belt.

Low-Solids Slurries for Clarification & Recovery: Batch Pressure Operation

Candle Filters and Pressure Plate Filters are installed for clarification and recovery applications from liquids with low solids content. These units offer full containment, fines removal in the 0.5 micron and finer range, and can be fully automated. The candle filters are vertical candles while the pressure plate filters are horizontal plates. The major difference between the two units depends on the cake structure that is formed. Some cakes are better handled in the horizontal and some in the vertical. Cake discharge is automatic either by gentle gas expansion of the filter sock in a candle filter or via plate vibration with gas-assist for the pressure plate filter.

Low-Solids Slurries for Clarification & Recovery: Conventional versus Contained Filter Presses

Conventional filter presses are open design with a series of vertically mounted plates. The Filter cloth is hung over each plate, extending from top to bottom, and is held in place by eyelets that fit over the cloth-pins on top of the plates. The plates are compressed by hydraulic closing pressure. Cake discharge can be manual or automatic.

Contained filter presses consist of circular filter plates that are sealed in a movable housing. The housing is sealed to 150 psig and full vacuum. The housing is then contained in an outer enclosure. The filter plate design allows for forward and reverse flow operations and can be automatically cleaned to less than 1 ppm residual cross contamination. Process steps can be pressure or vacuum and cake discharge is automatic using scraper knives.

CASE HISTORY EXAMPLES SHOWING "COMBINATION MECHANICAL SLURRY CONDITIONING AND FILTRATION"

The next sections review various actual installations showing the benefits of "combination mechanical slurry conditioning and filtration." These examples are included to illustrate creative idea-generation when examining a process problem.

Continuous Vacuum Filtration Followed by Candle Filtration

In this case, shown in Figure 4, the slurry contained a high solids loading but with a very small particle size distribution. Continuous vacuum filtration solved the first part of the process. The cake is very sticky with a cake thickness of 5 – 6 mm. The mother filtrate contained small, very fine solids. These solids were filtered, washed, and concentrated in candle filters and then sent back to the process.

Secondly, as the solids are sticky, the cloth wash contains solids. These solids, as above, are filtered and concentrated also in candle filters and then sent back to the process.

Candle Filtration Followed by Continuous Vacuum Filtration

In this case, shown in Figure 5, the candle filters operate to mechanically condition the slurry through thickening and concentrating up to 10% solids. The resulting slurry can then be economically and technically processed on a vacuum belt filter by vacuum filtration, cake washing and drying. The drying can be by vacuum, compression, blowing with hot or ambient-temperature gas or steaming. Applications for this approach occur in chemical plants as well as in coal gasification plants.

Candle Filtration Followed by Pressure Plate Filtration

This is the case described in the testing section "Typical Example of "Combination Mechanical Slurry Conditioning and Filtration" and shown in Figure 3. The process flow is shown in Figure 6.

Continuous Vacuum Filtration Followed by Contained Filter Press Filtration

In this process, there are three process reaction steps occurring which are being processed by only two types of filtration technologies. Figure 7 illustrates the process flow.

The first reaction requires vacuum filtration, cake washing and drying by vacuum only; the crystals are fragile and cannot be mechanically pressed. The continuous-indexing vacuum belt filter can have the pressing device installed but not operated.

The second step has more robust crystals and requires a lower moisture specification. The cake from the second reaction and is then processed on the same vacuum belt filter but, in this case, the cake is mechanically compressed with blowing for final drying.

The cake from the second step undergoes further processing, but, in this case, pressure filtration is the optimum solution. Due to the nature of the chemicals and solvents, a contained filter press for pressure filtration, cake washing and drying, by both blowing and vacuum is accomplished.

Continuous Pressure Filtration Followed by Candle Filtration

For this commodity chemical, with rates up to 60 tons of dry solids/hour, the initial mother liquor contains fines in the 1 – 5 micron size range. The process decision is to accept fines in the mother liquor rather than have a reduced rate with a tighter filter cloth. The fines are then recovered in the candle filters which can either produce a dry cake or concentrated slurry. The flow diagram is shown in Figure 8.

Continuous Vacuum Filtration (Two Variations)

In this case, similar to one of the previous examples, it was more efficient to use the same vacuum belt filter with two different variations and to conduct the initial cake washing in a separate agitated reslurry tank. Figure 9 shows this process.

Candle Filtration Followed by Conventional Filter Press

In some applications, a conventional filter press is the correct technology but the initial slurry feed is the problem. For these applications, mechanically conditioning of the slurry by concentrating candle filters is the optimum approach. The concentrating candle filters will increase the solids concentration in the slurry as well as reduce the volume of slurry that is to be treated by the filter press. Specialty chemical, chlor-alkali and coal gasification plants are typical examples. Figure 10 shows this flow process.

SUMMARY

This article discussed a new definition of combination filtration that will provide process engineers a framework for “idea-generation” when analyzing an operating bottleneck. The approach relies upon the slurry analysis and testing to uncover the “process symptom” and then develop a process solution called “combination mechanical slurry conditioning and filtration.” This article illustrated that the installation and combined use of filtration technologies, while higher in capital cost, will result in a more reliable operating process at the plant. In all cases, however, only through accurate lab and pilot testing, will the optimum filter selection be realized.

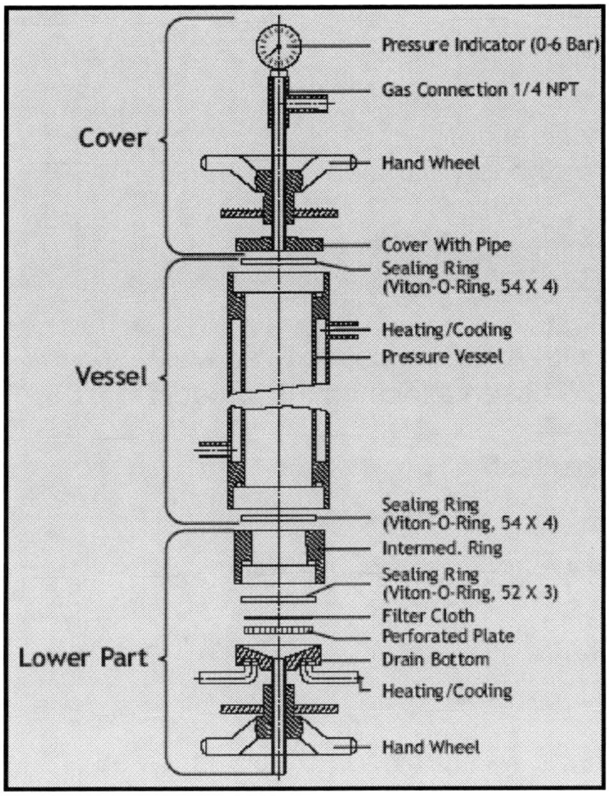

Figure 1: BHS Pocket Leaf Filter (PLF)

Customer :		Test Number:
Date :		Run #
	Filter Media	
	Suspension	
Filling	Volume of Slurry	
	Density of Slurry	
	% Solids in Feed	
	Temperature	
Filtration	Vacuum or Pressure	
	Volume of Filtrate	
	Time for Filtration	
	% Solids in Filtrate	
Wash 1	Wash Material	
	Pressure	
	Volume of Filtrate	
	Time for Filtration	
Wash 2	Wash Material	
	Pressure	
	Volume of Filtrate	
	Time for Filtration	
Drying	Pressure	
	Temperature	
	Flow Rate	
	Time for Drying	
	Pressing Pressure	
Cake	Weight	
	Thickness	
	% Residual Moisture	
	Dry Cake Weight	
	Cake Discharge OK?	

Figure 2: Data Collection Sheet for BHS Pocket Leaf Filter (PLF)

1. Test Purposes

The purposes of these tests were to:

1.1 Determine required filter area for the primary filtration.
1.2 Evaluate filtrate quality of both primary and secondary filtration.
1.3 Determine filter aid requirement for secondary filtration.
1.4 Evaluate washing efficiency and minimize amine loss in filter cake.

2. Test Methods

The BHS-Filtration Pocket Leaf Filter, jacketed with a 400 ml capacity and with 20 cm^2 filter area, was used to gather data and make observations on this product.

The following information was gathered during this test:

Primary Filtration Test Methods:

2.1 Variable speed peristaltic pump was used to conduct primary filtration tests.
2.2 Filtration Rate vs. Differential Pressure
2.3 Filtrate Quality vs. Filter Media

Secondary Filtration Test Methods:

2.4 Filtration pressure for secondary filtration tests was provided by compressed air.
2.5 Filtration Rate vs. Filter-Aid Concentration
2.6 Filtrate Quality vs. Filter-Aid Concentration
2.7 Filtration Rate vs. Filter Media
2.8 Filtrate Quality vs. Filter Media
2.9 Filter Cake Residual Amine (Filtrate Conductivity) vs. Volume Wash Water
2.10 Filter Cake Residual Moisture vs. Drying Air Usage

Figure 3: Typical Testing Program

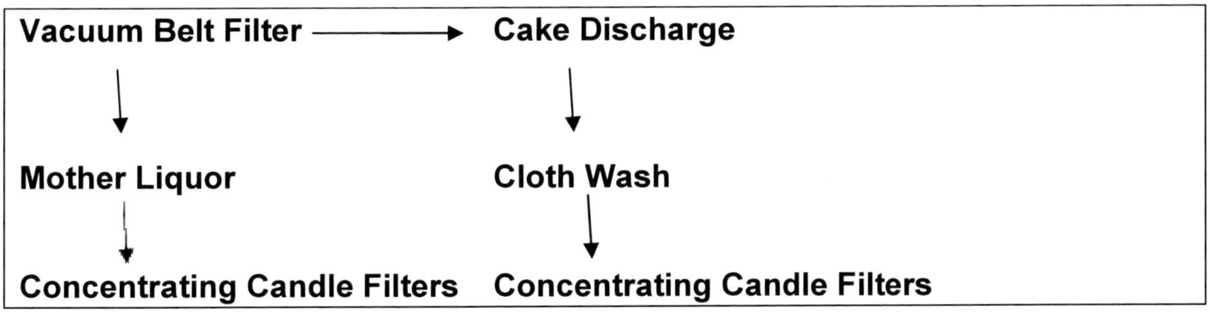

Figure 4: Continuous Vacuum Filtration Followed by Candle Filtration

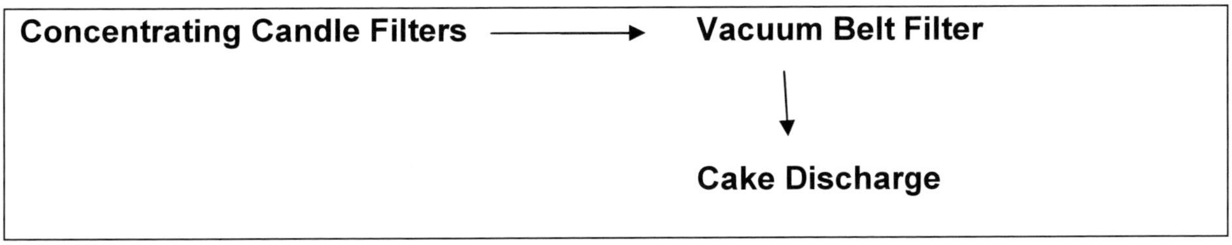

Figure 5: Concentrating Candle Filters Followed by Continuous Vacuum Filtration

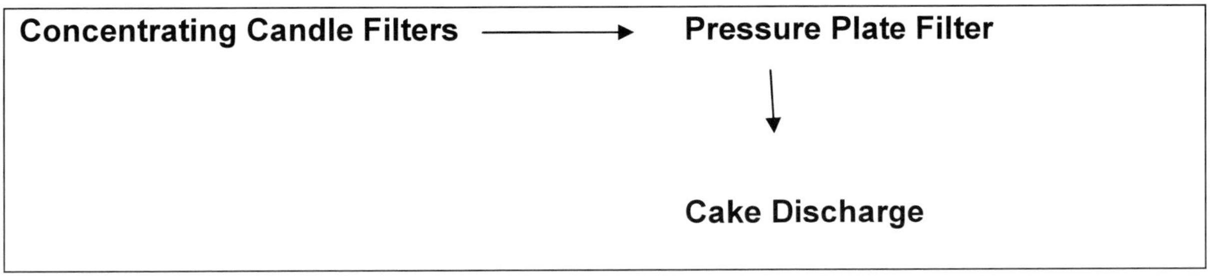

Figure 6: Concentrating Candle Filters Followed by Pressure Plate Batch Filtration

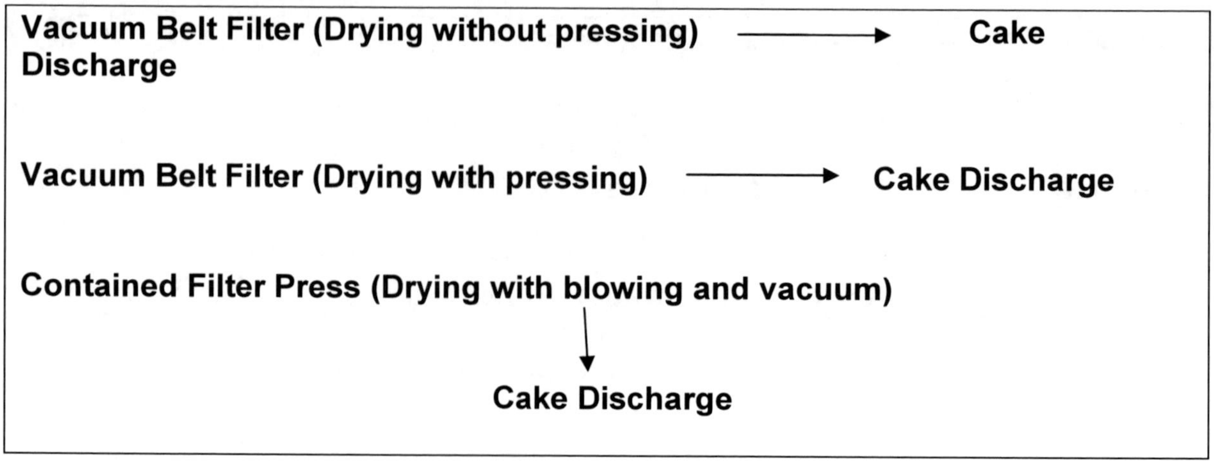

Figure 7: Continuous Vacuum Filtration (Two Variations) Followed by Contained filter Press

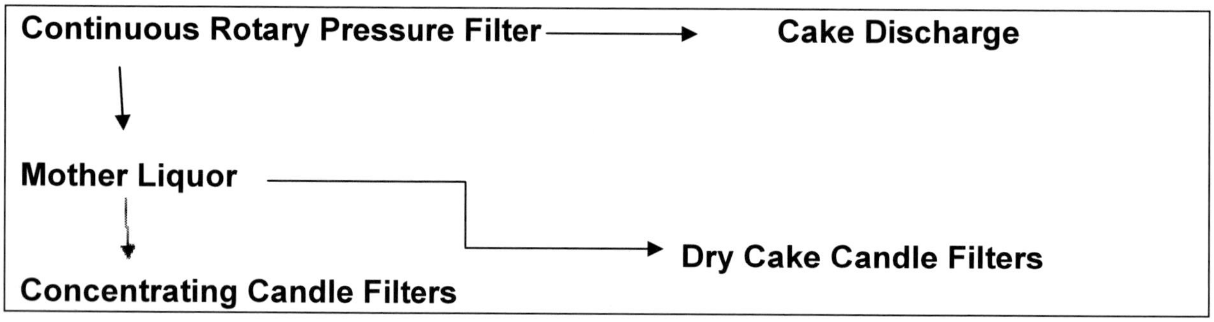

Figure 8: Continuous Rotary Pressure Filtration Followed by Candle Filtration

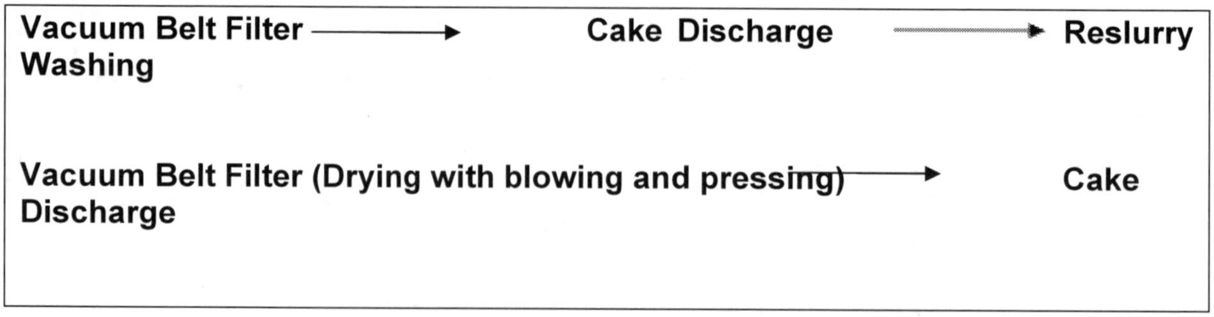

Figure 9: Continuous Vacuum Filtration (Two Variations)

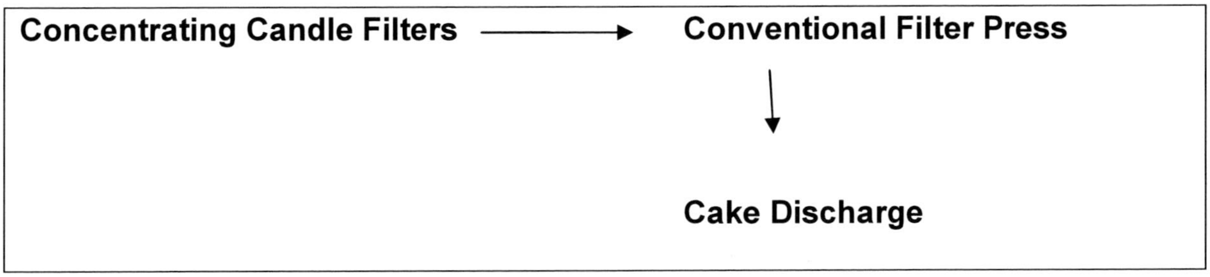

Figure 10: **Concentrating Candle Filters Followed by Conventional Filter Press Batch Filtration**

(53b) Prediction of Salt Effect on Vapor - Liquid Equilibrium of Methyl Acetate - Methanol System

Dhanalakshmi J., Sai P.S.T. and Balakrishnan A.R., Chemical engg, IIT Madras, chennai, India

The salting coefficient of salt in polar system (methyl acetate - methanol) is obtained by using scaled particle theory has been used. The scaled particle theory is a purely predictive method for evaluation of the salt effect on vapor - liquid equilibrium of methyl acetate - methanol at 1 atm. It requires only molecular parameters for calculating the salt effect. From scaled particle theory, the salt effect of polar molecule have examined in non-aqueous system. In the present study, the salting coefficient of polar molecule is predicted for various inorganic salts. The effect of cation and anion on the salting out phenomenon of methyl acetate - methanol system is reported. It is observed that salting coefficient is depending on the size and polarizability of the ions in non-aqueous solution. Therefore, the salting coefficient for a ternary system of methyl acetate - methanol - salt is well determined by scaled particle theory.

Keywords: methyl acetate, methanol, scaled particle theory, salting coefficient

(53d) Design of a Thermal Separation Process for Draw Solution Recovery in Forward Osmosis Desalination Plant

Tae-woo Kim, Department of Chemical and Biomolecular Engineering, Korea Advanced Institute of Science and Technology, Daejeon, South Korea, Young Kim, Energy Plant Research Division, Korea Institute of Machinery and Materials, Daejeon, South Korea, Woohyun Kim, Korea Advanced Institute of Science and Technology, Department of Chemical and Biomolecular Engineering , Daejeon, South Korea and Sunwon Park, Korea Advanced Institute of Science and Technology, Department of Chemical and Biomolecular Engineering, Daejeon, South Korea

The optimal design of a draw solution recovery process for a forward osmosis desalination plant is studied. The previous researches in forward osmosis (FO) desalination are mainly focused on the water permeability and salt rejection of semi-permeable membrane. However, relevant engineering issues such as recovery of draw solution and energy efficiency of process should be also resolved to commercialize a FO desalination process.

In this study, a continuous process is designed using a distillation column to separate draw solution and permeated water. The optimal design of the process is suggested achieving the minimum production cost while purity of product water is maintained at drinkable level. For case studies, ammonium bicarbonate, ammonium hydroxide and ethanol are employed as draw solutes which are selected by the screening method based on property database in our previous research. Each draw solute system is individually designed to handle redissolving or reconcentrating problems of each case.

(53e) Leaching of NdFeB Magent Scarp

Ho-Sung Yoon, Korea Institute of Geoscience and Mineral Resources, Daejeon, South Korea and Kyeong Woo Chung, Metal Recovery Department, Korea Institute of Geoscience and Mineral Resources, Daejeon, South Korea

Leaching of NdFeB magent scarp

Ho-sung Yoon*, Kyeong Woo Chung, Chul-ju Kim

Metal Recovery Dept., Korea Institute of Geoscience and Mineral Resources, Daejun 305-350, Korea

*E-mail: hsyoon@kigam.re.kr, Tel: +82-42-868-3611

Currently, large amounts of scarp form NdFeB magnet manufacture are stockpiled owing to the lack of a cost-effective processing method. Because of the growing new market for individual rare-earth metals such as Nd and owing to the likely continuing increase in demand for permanent magnets, scrap treatment will undoubtedly become an important alternative materials source. Development of a viable method to process NdFeB scarp will aid manufacturers economically and reduce the necessity of consuming rare-earth materials.

In the present study, Nd was recovered from NdFeB scrap. To dissolution Nd and Fe, scrap was oxidizing-roasted and leached in sulfuric acid solution. Nd and Fe was separated using double-salt precipitation method by the addition of sodium sulfate to leach liquor. Optimized temperature of oxidizing-roasting was 500°C for sintering magnet and 700°C for bonded magnet. When oxidizing-roasted NdFeB magnet was treated in 2M sulfuric acid at 50°C for 2hr with the pulp density of 15%, 99.4% of Nd and 95.7% of Fe was leached and recovered.

25 Years of Process Design + Manufacturing + R&D and Still Don't Know What I Want to Be When I Grow up

Ahmed J. Allawi, Innovation R&D, Shell Projects & Technology, Houston, TX

Tuesday, April 3, 2012: 8:35 AM
340A (Hilton of the Americas)

Abstract

The author spent most of his career in the Engineering and Construction industry (E&C) where he worked in process & environmental engineering, technology support, and project & department management. After 20 years in this industry, he took a leap of faith and joined one of his client companies to manage the manufacturing support functions at a major petrochemical complex. Two years later, he moved to his present role where he leads a regional Research & Development group within the same company. The author shares his first-hand experience of working in these numerous roles in three different industries. In addition, he discusses the value of advanced degrees in engineering and/or business in each of the industries in which he worked. The author's main goal for the paper is to provide practical career guidance to young engineers as they consider their options for professional development and advancement.

Introduction

"Why would a young engineer ever want to work for an Engineering & Construction company? No one ever listens to you before you hit 50!" That was the unsolicited advice of one of my former colleagues, which he offered me in the late 1980's. At that time, I was contemplating the option of leaving a job with a small start-up company to join one of the world's most respected Engineering & Construction (E&C) firms. Although I am glad I didn't heed his advice, I promised myself I would keep it fresh in my mind so that I can offer it someday as a (possibly) dissenting point of view.

Nearly a quarter of a century later, I find myself puzzled the scarcity of practical career guidance available for newly minted chemical engineers. Much as it was in my time, advice from older colleagues still plays a large role in forming both biased and unbiased opinions on employment options. While I don't claim to be unbiased, I think I can present my experience in a way that would help someone make more informed career choices. My hope is to provide the reader with the kind of insight that will generate more questions than answers. It is only by asking these questions and answering them for yourself can you make a truly informed choice.

Early Career Choices

As far back as I can remember, I've always wanted to be a chemical engineer. Growing up in the 1970s, I witnessed firsthand what the Arab oil embargo did not just to the economies of the Western world but also to the economy of Iraq, the country of my birth. The buzzwords of the day were "Synthetic Fuels", "Coal Gasification", and "Biomass". Big rewards awaited the companies and individuals who were the first to find the proverbial Holy Grail and produce low-cost fuel from inexpensive raw materials. This social economic climate had a profound influence on our profession. Chemical engineering enrollment swelled in the late 70s and early 80s, and by 1982 it reached a peak. At my alma mater, for example, the class of 1982 included no fewer than 220 chemical engineering graduates - almost a third of all graduating engineers. What we didn't know at that time was that we were in the midst of an artificial bubble. As soon as President Reagan stopped funding for the development of synthetic fuels, the chemical engineering world started a rapid decline. By the time I finished my undergraduate training two years later, my graduating class had shrunk by one half from the peak of 1982. Two years after that, the oil bust hit, and the size of the chemical engineering graduating class was down to a quarter of its peak. The disconcerting part was not in the drop in the supply of chemical engineers but rather the underlying drop in demand: In 1980, every chemical engineering graduate had multiple employment offers, but by 1984, only one third of my classmates could boast about an employment offer before graduation.

I am sharing this background because it's important to understand the context of the career advice and/or insight you might be receiving from your older colleagues. The chemical engineering graduates of the early and mid-80s are much like the survivors of the Great Depression: Our parents, grandparents, or even great-grandparents had to adapt their lifestyles in response to what they experienced in the Great Depression. In an eerily similar way, chemical engineering graduates of my generation had to shape their careers in response to the job market of the 1980s. Many got out of engineering altogether. A sizeable contingent elected to go directly into graduate school without even bothering to look for a job. Those who were fortunate enough to land a job with a good company at that time are probably still employed by the very same company today.

My circumstances then were little different than most of my colleagues. For one thing, I was on a student visa which disallowed employment after completing my undergraduate training. For another thing, I had a strong interest in pursuing postgraduate training - with the hope of eventually embarking on a career in process design at a major E&C company. Much as it is today, the optimal degree depended, to a large extent, on what the individual wanted to eventually do for a living: A bachelor's degree for operations support, a master's degree for process design, and a PhD for Research & Development. While the prospect of pursuing a PhD was very appealing to me, I could not convince myself that a PhD would be viewed as an asset in the E&C industry. I consider this to be one of the toughest career decisions I have ever had to make. On the one hand, I immensely enjoyed not just the research work I was doing but also the thrill of working with a world-renowned expert in a field that ignited my passion. On the other hand, I couldn't see myself doing the same line of work in a laboratory environment for a good part of the next four decades. I desperately wanted to be in a project world where I could see my ideas getting transformed from paper to steel and concrete. To me, that was what engineering was all about. Everything else had to be subordinated to that.

When I earned my master's degree in 1986, the engineering market was still in a deep trough. The days of multiple offers awaiting graduates were a distant memory by then. As chemical engineers began exploring career options outside the petrochemical world, non-traditional employers began picking up the slack in the job market. As a result, some of my colleagues found interesting jobs in electronics manufacturing, pharmaceuticals, environmental engineering, or equipment manufacturing. As for me, my most promising prospects were with a food manufacturer, a pharmaceutical company, and a start-up equipment manufacturer/environmental consulting firm. It was the last prospect that really thrilled me: A super-small, no-frills company, with 25 employees in two locations in, of all places, Connecticut. (Nothing against Connecticut – it's just that it's not exactly a Mecca for chemical engineers.) The company manufactured ethylene oxide emissions control equipment *and* provided environmental consulting and permitting services to manufacturing companies.

It would not be an exaggeration to claim that I learned more in the 27 months I spent with a small company than I did in the rest of my career since then. There was, literally, no aspect of operations that I did not experience first-hand while working for my first employer. My assignments included stints in direct sales & marketing, engineering design, environmental permitting, technical consulting, contracting & procurement, equipment fabrication, equipment installation, start-up, testing, and customer support. Less glamorous, but no less exciting, were my foray into support functions such as human resource management, business stationary design, computer repair, and, at least in one instance, sanitary engineering (the unclogging of a bathroom sink). It was in this environment that I learned how easy it is to lose money and how hard it is to make a profit. Personal accountability and ownership took on a different meaning: What I personally did (or didn't do) could impact whether the bank was going to honor our paychecks at the end of the month or let them bounce like a house of cards. And yes, there was one month where we couldn't meet our payroll needs, and several employees had to dip into their personal savings and credit cards to make ends meet.

As you can surmise from the above, working for a small start-up company can be highly rewarding – both personally and professionally. These rewards, however, come at the expense of higher risk: The nature and volume of the work can be highly erratic, the growth opportunities can be severely limited, and the company can go bankrupt. If you can tolerate these risk factors, especially early in your career, the experience you gain from working with a small company are worth serious consideration.

Jumping Ship for the First Time

I mentioned earlier that leaving graduate school was one of the toughest decisions I made in my career. The toughest, however, was leaving my first employer. My risk tolerance was high, and I thoroughly enjoyed the work, the atmosphere, and the company of my coworkers. However, I realized that my growth opportunities were coming to a screeching halt. Fortunately, I came to this realization as the market for chemical engineers was warming up in the late 1980's. The oil bust of the mid 1980's was being replaced by an ethylene boom, and virtually all major E&C companies were actively looking for chemical engineers once again. The timing could not have been more perfect for me to enter the major projects world. I actively sought and then landed a role with a technology-based E&C company.

On January 16, 1989, I was one of 30 new-hires to start working for my new employer, a major E&C company headquartered in Houston. What I found ironic was that more new-hires started with me that day than the entire employee population of my previous employer. Gone were the days of writing my own proposals and purchase orders, and in were the days of strictly following procedures and operating manuals. I had to keep reminding myself that I didn't join an E&C firm because I enjoyed the bureaucracy of a large company. Rather, I was there to be part of an engineering team that will design and bring to life one of the most complex machines ever built by our profession: An ethylene plant. The bureaucracy and the procedures were simply means to an end.

The scale of a major engineering project is hard for an outsider to imagine: It takes upward of a thousand engineering work-hours to fully design and integrate a single piece of equipment into a chemical plant. Think back to your process design class in college where you may have prepared a preliminary process design for a simple distillation system. The system may have consisted of a distillation column with reboiler, condenser, reflux accumulator, and three pairs of pumps – 10 pieces of equipment in total. For these 10 pieces of equipment, an E&C company will spend approximately 10,000 engineering work-hours to develop a fully detailed specification package. Now imagine the scope of a major grass-roots refinery project, where the equipment count could be in the *thousands*. It is not unusual for a major E&C project to tie up *hundreds* of engineers and designers for *several years* from the start of the project to the finish. It is precisely this complexity that makes this type of work so exciting and challenging.

Chemical process engineering in project execution involves developing, among other things, heat & material balances, process equipment datasheets, process flow diagrams (PFDs), and Piping & Instrumentation Diagrams (P&IDs) for the plant. As important as we think our role is in project execution, chemical process engineering typically amounts to less than 15% of the total engineering effort on a project. The rest of the engineering work involves piping design, civil/structural engineering, control systems engineering, electrical engineering, machinery, equipment design, and environmental support. When viewed from this angle, the required oversight and bureaucracy to coordinate all of these activities in a consistent and efficient manner becomes clear. That is basically what a project engineer does on a major project.

The anatomy of an E&C project as presented above partially explains why the various engineering disciplines are highly compartmentalized. Process engineers may have a lead role in conducting process simulations and developing heat & material balances for the plant. However, they only play an advisory role for the other engineering disciplines. Similarly, equipment engineers may play a lead role in specifying engineered equipment (such as compressors, pumps, and exchangers), but they play only an advisory role in process design. The implications of this on career development are significant: An engineer at an E&C company will typically become a specialist in one discipline and may never get the opportunity to work in different technical disciplines. Coming from a background where I was used to doing many different things, this was a very hard pill to swallow. Fortunately, my supervisors were very supportive of my need to work in different fields: For many years, I was one of a handful of engineers who could claim "dual citizenship", one in process engineering and one in environmental engineering. The vast majority of my colleagues, especially those who grew up in the E&C world, were in "single citizenship" roles, and they didn't feel the need to venture too far from their Mother Ships. I bring this point up because there is an important attribute of the E&C world that is frequently overlooked: discipline longevity. There are no expectations for a newcomer to move from one discipline to another unless that happens to be what the individual wants. For example, a person who starts out as a process engineer could very well remain as a process engineer for his/her entire career without ever working in a different discipline. This is in stark contrast to the owner/operator world, where an individual is *expected* to move to different departments, disciplines, or groups as a broadening opportunity (more on that later).

With discipline longevity comes an extremely strong sense of belonging in the E&C industry. Members of one group or discipline could easily work together for decades on a myriad of projects without ever losing touch with each other. This results in tightly-knit networks and strong bonds among members of the same group or discipline. Nothing exemplified this better for me than by looking around the table at my farewell lunch: I had known and worked with each individual around the table either from the time I joined the company, or from the time they joined the company, whichever came later. There were no "outsiders" who came into the group through osmosis or diffusion from other groups; it was and will always be a *mass transfer limited operation.*

Discipline longevity also explains why most employees of E&C companies don't jump ship to owner-operators when the opportunity arises. In general, the E&C industry is considered to be more volatile than owner-operators when it comes to job security. Rarely is there a perfect balance between workload and headcount. As a result, the E&C industry goes through relatively frequent cycles of growth and contraction. The impact on the affected individuals could be devastating. With this omnipresent risk, one would expect that most employees of E&C companies would leap at the opportunity to join a more secure employer such as an owner-operator. A colleague of mine explained it best when he said the reason why he hadn't sought employment with one of his client firms is because "*I didn't want to have a different job in a different location every two years*". At the risk of over-generalization, the E&C world values depth and discipline focus whereas the owner-operator world values breadth and multi-discipline.

In no way am I implying that it's not possible for a chemical engineer to work in different *functions* at an E&C company. Personally, I had the fortune of being able to work in many different areas including process design & systems engineering, environmental engineering, project engineering, project management, and technology support. As my focus and interest shifted from purely technical roles to managerial roles, I took on the responsibility of leading a process technology group and then became a resource manager for the Project Definition department. As broad as these assignments sound, however, they were not that far from process engineering, which is what I was hired to do back in 1989. Even my office in 2006 was only a few feet away from where it had been 16 years earlier.

From E&C to an Owner-Operator

The E&C industry underwent a significant transformation in the early- to mid- 2000's. Much of the work shifted from North America to the Middle and Far East. The chemical industry was heading for a downturn, but the refining, LNG, and gas monetization sectors were on the rise. Suddenly, a $300 million project was no longer considered sizeable. Capital projects in the multi-billion dollar range were becoming more prevalent. The impact of this paradigm shift on the E&C industry was profound: Companies that were chemical technology-based began scrambling for engineering resources to staff the super-mega-projects for the refining and gas processing industries. For me, this shift meant that I had two options: Either align my career along a project execution pathway or look for a chemicals technology-based opportunity in a different industry. For various reasons, I opted for the latter.

Over the course of my career in E&C, I was fortunate to have developed close working relationships with a few of my clients. Through my contacts, I learned of an opportunity to manage a field technology group at a manufacturing complex. The technology component of the job certainly appealed to me, but I wasn't sure about the manufacturing support aspects. Clearly, it was a new area for me, and while I was happy to try it out, I didn't know if I would enjoy doing this line of work for a long period of time. Within an owner-operator organization, however, I knew that the range of career options was virtually limitless. That being the case, I was willing to take the chance that if manufacturing was not a good fit for me, I could seek a re-direction at some point in the future.

It should come as no surprise to anyone that operations or, more specifically, manufacturing support is about as different from engineering design as can be. Operation support engineers are like fire-fighters: They have to be tactical, comfortable working with incomplete data, able to think on their feet, naturally intuitive, thrive on adrenaline, and have an exclusive focus on getting broken things back on track. Most of their activities are short-lived, and long-term planning only goes as far as the next plant turnaround. Their time is not their own to manage: When the plant shuts down unexpectedly, it's all hands-on-deck regardless of the time of day or night, or what other activities had been previously scheduled or agreed-upon.

Process design engineers who move into plant support roles tend to get frustrated by the constantly shifting landscape of the production world. On the opposite end of the spectrum, production support engineers who move into the engineering design world tend to get frustrated with the level of detail and methodical approach required to develop a complete design package. Unlike the E&C industry where a process engineer doesn't venture too far from his/her discipline, a process engineer with an owner-operator has the luxury of being able to experience both ends of the spectrum. Rotation of developing engineers into operations support roles is not merely encouraged in an owner-operator organizations, it is a requirement for advancement in many cases. The idea, as I pointed out earlier, is for owner-operator engineers to develop breadth and become multi-disciplined as they advance in their careers. Many owner-operators, in fact, have highly structured programs that force new engineers to go through three or more rotations in different disciplines before allowing them to select their career paths. At my company, typical rotations include stints in capital projects, operations support, control systems engineering, and other disciplines. After the developing engineers complete these rotations, they are free to seek assignments in any field of their choice.

Although I had no intention of spending the rest of my career in operations support, I didn't realize how quickly the next assignment opportunity would emerge. A little over two years into my tenure, a managerial position with the R&D group sparked my interest. The nature of the work had much in common with what I had experienced in the E&C industry: In both settings, project time spans are typically of the order of a few years, and project execution is always methodical and sequential. In E&C, a project starts with a conceptual design and then proceeds through project definition, detailed engineering, procurement, and construction. In R&D, we typically describe a project using the "4D" model: Discover, Develop, Demonstrate, and Deploy. Neither setting affords you the luxury of skipping a step without assuming undue risk.

The 4D model concisely describes what a process development engineer experiences in an R&D environment. The process begins with a highly inquisitive individual discovering a new catalyst, a new molecule, or a new pathway to a product. A process evaluations engineer then conducts a preliminary study to determine if the novelty has technical and economic merit. If deemed feasible and strategic, the concept is tested in bench-scale rigs to establish an operational envelope and validate the technical feasibility of the critical steps. The experimental results from the bench-scale rigs are then used to develop a design for a suitably integrated pilot plant. The piloting step is, perhaps, the most important phase of an R&D project. Even though a project can be deemed infeasible at any phase of execution, it is the piloting phase that provides the most useful data to support a go/no-go decision. Although this phase is necessary, it is not always sufficient to assure successful commercialization of the concept. In many cases, a fully-integrated demonstration unit (semi-works scale) may be required for a variety of reasons. These may include verifying the scale-up approach, producing commercial quantities of a new product for customer testing, or providing assurance to investors that the project will, indeed, meet its financial objectives. If the demonstration unit meets all of its objectives, then the commercialization work can proceed.

Not every R&D project is successfully commercialized. Most projects, in fact, do not make it out of the "development funnel". This implies that development engineers may work for years on a myriad of projects without ever experiencing the commercialization phase in person. In an extreme case, a colleague who retired after a quarter of a century in this line of work casually mentioned at his retirement party that he had never worked on a project that was commercialized. For those of us who like to see their work in steel and concrete, this aspect of the R&D world is sure to bring some periodic disappointments. Conversely, an engineer who develops a project from concept all the way to commercialization will achieve a state of gratification that few in our profession will ever experience.

To MBA or Not to MBA

A highly "objective" colleague of mine once proclaimed: *Anything a mechanical engineer can do, a chemical engineer can do even better.* At the risk of plagiarizing my colleague's thoughts, I can add a highly objective corollary to his theorem: *Anything that an MBA can do, a holder of a graduate degree in engineering can do even better.* So why would a member of our profession willingly choose to dilute his/her analytical capability by pursuing a business degree?

If you cannot articulate *exactly* what it is that you will be doing with your MBA degree, you shouldn't even think about going to business school. The best advice on the subject can be found in a book titled "Business School Confidential" [1]. If you are thinking about going to business school, you owe it to yourself to read this book cover-to-cover before you make up your mind.

If you are pursuing an MBA to get into a different industry such as finance or management consulting, then good for you! That's where the degree would be worth your investment, and it will give you opportunities that would otherwise be unavailable to you as an engineer. Just make sure that you attend a top-rated school to get the best options after graduation.

If you want to stay with your current employer and believe that an MBA will give you different or better career options, think twice before taking the plunge. First, you need to make sure that the job you want does, in fact, require an MBA. Would your employer select you for this job because of your known track record or because of your new and shiny diploma? Second, think about whether an MBA is a good fit with the business model of your employer. The simpler the business model, the less valuable is an MBA degree to an employer. For instance, a mom-and-pop type business with one revenue stream and one cost stream has very little to optimize. On the other hand, a multi-national conglomerate with thousands of revenue streams will need hordes of MBAs to manage its complex operations. The E&C industry is certainly much closer to the simple end of the business model spectrum than, say, a major petrochemical company. This is not to say that an MBA wouldn't be valuable to an E&C company; Just that the opportunities for an MBA with an E&C company would be more limited than those in an owner-operator organization.

Finally, your interest in an MBA could very well be driven by a desire to learn the tools-of-the-trade rather than an interest in changing careers. In this case, you should pursue your degree on a part-time basis with a local school. This sounds a lot easier than it really is, especially if you have a family or an unpredictable work schedule. On the plus side, the distraction offered by part-time studies could be a welcomed change from the rigors of your day-to-day responsibilities at work.

I'll close this section by rephrasing my earlier tongue-in-cheek comment in a more serious manner: In my humble opinion, a graduate degree in science or engineering will enhance your analytical skills much more effectively than an MBA. You already have all the necessary skills to learn, *through self-study*, everything you need to know about finance, marketing, strategy, and organizational behavior. Have you ever seen an MBA learn engineering through self-study?

Closure

If your goal in reading this short paper was to identify the perfect job for yourself, I hope you're utterly disappointed by what you've learned from me! My goal was not to point you to Utopia but to share with you some of the realities of what it's like to work in E&C, manufacturing, or R&D. The best fit is clearly a personal choice. But more importantly, even a not-so-good fit can offer invaluable growth and learning opportunities - if you can keep an open mind and maintain a positive attitude.

Reference

Miller, R.H., and Koegler, K.F. Business School Confidential. St. Martin's Press, New York, 2003.

(69b) Process Evaluations & Manufacturing Support: From Concept to Commercial

Todd A. Whittemore, Team Leader - Process Evaluations , Shell International Exploration & Production Inc, Houston, TX

Through personal experience with projects ranging from minor debottlenecks to major grassroots petrochemical projects executed globally, Todd will share the importance of project development integrating product slate, process flowsheet, Licensor selection, partnering, option development, and techno-economic evaluation. Coupled with the initial selection of the overall project scope, Todd will discuss the importance of building a winning organization comprised of a delicately balanced blend of experience with strong local membership. Todd will use numerous real life examples highlighting both successful applications as well as major learning experiences to display the complexity involved to bring a mega-project from concept to commercial success.

Planning and Developing a Career in Process Development

Jonathan H. Worstell

Abstract

The first ten years of a chemical engineer's career are critical to the path that career takes, to the success achieved during that career, and to the happiness, fulfillment, and respect obtained from that career. This talk presents the questions which must be answered in the early years of a chemical engineering career. Some of those questions are:

- Am I best suited for a technical or management career?
- Do I need an MBA?
- How do I develop a technical career?
- How do I develop a management career?
- Should I get plant experience?
- Should I earn a graduate degree while working?
- How do I learn to be a leader?
- Where do I get leadership training?

These questions and many others will be discussed from the insight gathered during a thirty year career in Process Development.

Introduction

You can study a particular subject at many different levels. For example, consider biology. You can study biology at the molecular level; that is, inside a single cell. Molecular biology and biochemistry study life at this level. Identification of the structural features of a cell is done at this level also. This level is highly technical, being based on the laws of chemistry and physics. At this level, you study the forces impacting individual molecules and structures. The next level is cellular physiology --- the study of how the molecules and structural features of a cell interact to produce change and energy. This level investigates transformation --- it studies the flow of atoms and molecules into the cell, how the cell uses them to produce various molecules, and the flow of proteins and by-products from the cell. At this level, you study the forces impacting the coherent, unitized cell. The highest level studies how cell aggregates interact with each other; i.e., it studies living animals and plants. At this level, you study the natural history of animals and plants. Natural history includes the study of physiology, individual behavior; i.e., psychology, and social behavior; i.e., sociology. At this level, you study the forces impacting animal and plant populations.

A similar analogy can be made with the study of economics. You could study the economics of a single process or a single product in isolation from the national economy as whole. Such a study is called Process/Product Economics. Process/Product Economics is analogous to molecular biology and biochemistry. Process/Product Economics requires detailed technical knowledge of the process/product and applies the principles of chemistry and physics to the needs of humankind. The next level of economics studies how multiple processes/products forms individual businesses. This level is called Microeconomics and it is analogous to cellular physiology. Finally, you can study how a national economy functions. A national economy is an aggregate of businesses and business activities. This level is called Macroeconomics and is analogous to investigating the natural history of animals and plants.

Thus, the private sector economy has three levels: Process/Product Economics, Microeconomics, and Macroeconomics. Process/Product Economics is the lowest or simplest level. This level identifies trends impacting individual processes and individual products. Microeconomics is the next higher level --- it identifies trends impacting groups of manufacturing processes, product groups, and business centers. Macroeconomics is the highest level. Macroeconomics identifies trends impacting national economies and industrial groups. Figure 1 shows these generalized levels of the private sector economy.

The organization of a commercial entity; i.e., a company or corporation mirrors the organization of the private sector economy. The lowest level of a commercial organization is the Entry Level. Newly graduated engineers and scientists enter the commercial organization at this level. At this level, employees accept responsibility for learning the details of an assigned process or product. Thus, they become responsible for identifying trends impacting individual processes or products. This level corresponds to the Process or Product Economics level of the private sector economy. The next higher level in a commercial organization is the Senior Level. At this level, employees are responsible for identifying trends impacting business centers, groups of manufacturing processes, and product groups. This level is equivalent to the

Microeconomic level of the private sector economy. Organization Leadership comprises the highest level of a commercial organization. Organization Leadership is responsible for identifying trends in national economies and industrial groups. Thus, it mirrors the Macroeconomic level of the private sector economy. Figure 2 shows these generalized levels for a commercial organization.

Most newly graduated engineers and scientists have little knowledge of how the private sector economy is organized or of how a commercial entity is organized. However, such knowledge is imperative when planning and developing a career in the private sector.

Career Development Generalizations

After graduating from university, most of us work forty to forty-five years, then retire. Those years are filled with work and comprise our "career". Prior to those forty or forty-five years; i.e., during our educational years, we in general receive no information about career planning, career development, or even industrial or economic organization. Until you understand industrial and economic organization, it is difficult to plan your career and without a career plan, you cannot efficiently develop your career.

This laissez faire attitude starts in high school. When asked what we want to be, we answer an engineer or scientist, thus our career choice. However, beyond choosing to attend U of H or UT Austin or Texas A&M, we make no plans for developing our careers. When at university, career planning means following the required course sequence, thereby ensuring graduation ... career development means studying hard enough to pass each course. When interviewing for our first job, we tend not to consider career development: we are concerned with getting a job offer --- and our parents are just as concerned. Once hired, we fall into believing that the company will plan, manage, and develop our career, which may have been true in the past, but is no longer true today. Today, embracing that belief is perilous to career development. Today, you must plan, manage, and develop your career yourself. But then: who is better qualified to plan, manage, and develop your career --- you or someone in Human Resources?

Figure 2 provides guidance for planning and developing a career in a commercial organization in the private sector economy. Figure 2 shows that from the Entry Level, you proceed to the Senior Level and from there to Organization Leadership. Thus, a given career path can be traced as an arc inside such an organizational triangle. Figure 3 depicts several career paths. The length of an arc represents the relative time encompassed by a career. The flatter the arc, the faster a career develops.

To advance from the Entry Level to the Senior Level, you must prepare yourself to function at the Senior Level while performing the functions of the Entry Level. And, to advance from the Senior Level to Organization Leadership, you must prepare yourself to function as a Leader while performing the functions of the Senior Level. Therefore, you must formulate a plan that gives direction to your career. Unfortunately, most engineers and scientists do not put much thought into providing their career with direction: they "go with the flow", letting external forces set the direction and pace of their career. After forty or forty-five years, most finish their "career" vaguely dissatisfied, if not disappointed.

There are many paths from Entry to Organization Leadership. In a commercial organization utilizing technology to produce a product or a variety of products, you can proceed from Entry to Organization Leadership via a career in the Business Group, the Manufacturing Group, or the Technical Group. Management and Staff comprise each such Group, as shown in Figure 4. We are all familiar with the Business Group --- it is responsible for the supply and delivery of products to customers at the highest possible profit margin. The Manufacturing Group oversees the conversion of chemical and material inputs to product at the lowest possible manufacturing cost. The Technology Group support and improve current production processes as well as devise new processes for manufacturing product.

In small commercial organizations, many of the functions shown in Figure 4 overlap; thus, you may find yourself functioning in the Business Group and the Technology Group or some other combination of functions. Doing so broadens your organizational outlook. In medium to large commercial organizations, the functions shown in Figure 4 are separated. Therefore, you must spend time in each function to gain the necessarily broad insight into the organization required of Organization

Leadership. In general, you begin your employment in one part of the organization, then fill an assignment in a different part of the organization. Upon completion of this assignment, you may return to your original part of the organization and take a second assignment in the future or you may immediately take that second assignment immediately upon completion of the first assignment.

Newly graduated B.S. engineers are generally placed in the Manufacturing Group of an organization. Similarly, newly graduated B.S. scientists are placed in the quality assurance and environmental groups at manufacturing facilities. Most graduating M.S. and Ph.D. engineers and scientists are placed in the Technology Group of an organization. The technology Group may or may not be located at a manufacturing facility. If your Entry point is the Manufacturing Group, then you will want to obtain assignments in the Technology Group and the Business Group. If your Entry point is the Technology Group, then you will want to obtain assignments in the Manufacturing and Business Groups.

With the generalizations depicted in Figures 1 through 4, you can begin planning and developing your career.

Career Development
Career Planning

To develop a career plan, you must reserve a morning or afternoon so you can sit and cogitate undisturbed. Saturday is generally the best day for career planning: it is your time, not company time, thus you control it. If you are single, you can sit at your kitchen table and cogitate. If married, it is best to remove yourself to a quiet corner of the local public library or to your company's library or to a quiet table at your favorite coffee shop. Be sure to have an adequate supply of blank paper and a writing instrument so you can record your goals and document your plan for achieving those goals. With today's technology, the same can be done using a laptop computer.

To establish your goals, you must first decide where you want to be in twenty years. You have to know where you want to be in twenty years in order to know where you ought to be in ten years. And, you have to know where you want to be in ten years to know where you ought to be in five years. Thus, you will have three sets of goals: a five year set, a ten year set, and a twenty year set. You have to have all these goals because advancement from the Entry Level to the Senior Level requires you to prepare yourself for the Senior Level while fulfilling the functions of the Entry Level and advancement from the Senior Level to Organization Leadership requires you to prepare yourself to be a Leader while fulfilling the functions of the Senior Level.

Career Development
Technical or Business

Two career paths exist for engineers and scientists in most medium- and large-size companies. Those career paths are: technical-oriented and business-oriented. Both career paths provide growth for the individual engineer or scientist. The technical career path allows the engineer or scientist to develop technical skills in an ever-broadening economic context --- beginning at the Entry Level, progressing through the Senior Level, then concluding at Organization Leadership. The business career path allows the engineer or scientist to develop and apply business skills in the same ever-broadening economic context.

These career paths coincide upon hiring and during the first few years of a career, as shown in Figure 5. Thus, the first choice you face as a newly graduated engineer or scientist is which career path to follow. This decision should be made within the first three years of starting your career because subsequent career planning depends upon your choice. In fact, a career does not begin until you develop a plan. Such a plan provides direction and goals for your personal development. Without a career plan, you respond to the vagaries of your work and social environments. In other words, "you go with the flow" --- you do not direct the flow of your career.

Before making the choice between a technical-oriented career and a business-oriented career, you must analyze yourself. You must determine what type of person you are. Are you

- quiet or vocal?
- subdued or splashy?
- retiring or social?
- introverted or extroverted?

- passive or dominating?
- logical or intuitive?
- contemplative or abrupt?
- controlling or relaxed?

To assess your personality, take a blank sheet of paper and half it lengthwise with a vertical line. Label the left side "Strengths" and the right side "Weaknesses", then sit and review the above pairings and other personality traits you deem important. Record your results on your halved sheet of paper.

Remember, we humans tend to see what we want to see when viewing ourselves. Thus, do not be shy about asking others for input regarding this assessment. Ask your parents for input: they have only been observing, studying, and analyzing you for twenty-some years. Also, ask your friends for input for this assessment. When making such requests, remember that their "eye" is different from your "eye"; in most cases, they see you more clearly than you see yourself. Also, remember that your purpose is to match your personality with a career path: success, thus personal happiness, devolves from making the best match of personality and career. If you are extroverted, like meeting strangers, and enjoy interacting with all kinds of people, then you have the makings of a successful business person. On the other hand, if you are introverted and do not like meeting strangers or making small talk with others, and possess a modest understanding of mathematics, then you should consider a technically-oriented career.

With the above information in hand, you can make the most important decision of your career: whether to develop a technically-oriented career or a business-oriented career.

Career Development
Technology Management or Technical Resource Management

A career in Process Development means choosing the technically-oriented career. Note, however, that there are two aspects to Process Development: Technology Management and Technical Resource Management. Technology Management means you will be determining the current and future success of your employer. In other words, you will be improving current processes, thereby ensuring that product is manufactured at the lowest possible cost. You will also be identifying, developing, and implementing the next generation processes, thereby again ensuring that product is manufactured at the lowest possible cost. Technical Resource Management provides the technology staff with the resources necessary to support and improve current manufacturing processes and to develop new manufacturing processes.

If you do not like coaching people and if disciplining or firing people is distasteful to you, then you should consider pursuing a career in Technology Management. However, if you enjoy coaching people and if the thought of disciplining or firing people does not keep you awake at night, then you might consider a career in Technical Resource Management.

Invention, Innovation, and Sales

What Is Process Development?

There are many kinds of people inhabiting this world; three of them are Inventors, Innovators, and Sellers. An Inventor is one who comes upon, discovers a new concept, experimental fact, or way to assembly of equipment in a wholly different, beneficial manner. An Innovator is one who converts an invention into a commercial process or product and a Seller is one who induces, moves the public to buy the commercialized invention. For example, Benjamin Franklin proposed the concept of electricity and suggested that it flowed in a positive stream and a negative stream. Michael Faraday expanded Franklin's ideas and concepts in word pictures. Thomas Edison utilized Franklin's and Faraday's ideas and concepts to develop many instruments and machines to convert electricity into work. Those machines were manufactured and marketed by companies such as Western Electric and General Electric in the United States and Siemens in Europe. In this example, Franklin and Faraday are Inventors, Edison is an Innovator and the sales staffs of Western Electric and General Electric are Sellers.

In chemical terms, petroleum "cracking" provides a similar example of process development. Benjamin Silliman discovered in 1855 that the application of intense heat to crude petroleum decomposed; i.e., “cracked”, the petroleum, thereby yielding volatile products. Between 1909 and 1913, William Burton of Standard Oil Company (Indiana developed a process utilizing Silliman’s discovery and commercialized it at the Whiting Refining. Standard Oil Company (Indiana) then licensed the Burton process to other petroleum refiners. Silliman was an Inventor, Burton was an Innovator, and Standard Oil Company (Indiana) was Sellers. Thus, chemical process development is innovation.

Process Development
Job Description

Process development is innovation. It takes an invention and formulates, then develops a process that produces commercially viable product. Consider this example. The Greatly Expensive Chemical Company, a competitor of your employer, produces molecule C, a molecule much desired by the market, via the reaction

$$A + B \rightarrow C$$

Let us assume their process is complicated; i.e., expensive. And, due to the market’s ravenous desire for C, your competitor gets their asking price, thereby making a tidy profit. This niche market needs competition. With considerable effort, the chemists at your commercial organization find that reactants X and Y yield C via the reaction

$$X + Y \rightarrow C$$

X and Y are not onerously expensive and you do a quick economic evaluation of an ideal process that suggests your employer can market C at a significantly reduced price, thereby gaining market share from Greatly Expensive Chemical Company. Thus, your employer decides to develop a process based on reactants X and Y.

The issues that you must resolve to produce C are

- TSCA registration of X and Y;
- inherent safety of X and Y;

- inherent safety of C;
- inherent safety of proposed process;
- pressure requirements;
- timeline using project management software;
- solvent usage;
- inherent safety of solvent;
- solvent purity;
- heat required or produced by the reaction;
- analytical methods;
- homogeneous catalyst;
- heterogeneous catalyst;
- reaction mechanism and reaction order;
- temperature dependence of reaction;
- by-product formation per order of reactant addition;
- temperature dependence of by-product formation;
- thermodynamic data;
- reaction engineering data;
- type of reactor --- batch, semi-batch, CSTR, bubble column, plug flow column;
- by-product formation per reactor type;
- by-product formation per extent of reaction;
- available laboratory reactors;
- available pilot plant reactors;
- process design options;
- separation of product from solvent, feeds;
- by-product (waste) management;
- solvent recycle;
- feed recycle;
- crystallization;
- drying;

- bulk handling;
- feed purity;
- solvent purity;
- product purity;
- PMN notice;
- MSDS;
- process engineering data;
- market sample;
- feed storage stability;
- solvent storage stability;
- product storage stability;
- semi-works availability --- use of contract manufacturing facility option;
- shipping for feed, solvent, and product --- tank wagon, rail tank car, barge.

This list is not comprehensive, rather it is suggestive. What this list indicates is the breadth of knowledge you need successfully this and other processes. Thus, Process Development is a multi-functional task.

Process Development
Required Knowledge Base

Pursuing a career in Process Development requires a thorough knowledge of the "first" principles of Physics and Chemistry and the application of those principles to everyday life and the real world. Thus, to maximize your career in Process Development, you need a sound grounding in Fluid Dynamics, Stage-wise Unit Operations (Absorption, Extraction, and Distillation), Heat Transfer, and Chemical Reactor Engineering, which means taking graduate-level courses. You can take these courses as a full-time student, thereby earning a Masters or Doctorate, then search for a Process Development position, or you can take these courses part-time, while working at a manufacturing facility or at a technology; i.e., R&D, facility. From a Process Development viewpoint, the most important point of these courses is the boundary layer concept --- the concept of a stagnant layer or film on both sides of a gas-liquid interface or a liquid-liquid interface or in the fluid phase adjacent to a solid surface, such as a catalyst surface or a heat transfer surface. This concept allows you to visualize the movement of mass from one phase to another phase.

You also need a solid foundation in Chemical Thermodynamics. First, Chemical Thermodynamics will tell you whether a process or process modification is possible. Second, in conjunction with Heat Transfer, it provides information about the potential safety hazards of a process or process modification.

The above courses provide the technical skills for developing a new process or for modifying a current process, but at some point, the process has to be transferred to a production facility. When that transfer begins, you will need a good understanding of how such a transfer impacts the environmental permit for the plant and the safe operation of the plant. It thus behooves you to take environmental courses and a process safety course to enhance your knowledge concerning these plant issues. Universities offer such courses, as do professional societies during national conferences. Remember; you do not have to be an expert in these areas, but you do need to know what such experts worry about and you need to know the vocabulary they use to express their worries.

Career Development
Twenty Year Goals --- Technical Fellow

Establish your twenty year goals first. Doing so provides direction to your career. Your twenty year goals also require you making your first career-oriented decision: to pursue a career in Technology Management or to pursue a career in Technical Resource Management. If you want to be in Organization Leadership, for example, become the Vice President of Technology or R&D, then you will tread the Technical Resource Management career path. If you want to be a Technical Fellow in twenty years, then you will tread the Technology Management career path.

Remember: in Organization Leadership, you identify trends that impact national economies and industrial groups. This level involves leadership --- functions through leadership. Managing and leading are different activities: managing involves transmitting direction from Organization Leadership; i.e., telling employees to follow that direction and monitoring employees to ensure that they are following the prescribed direction. Leading promulgates a vision and achieves influence through example. Leading is not telling, monitoring, supervising.

Both Technical Fellow and Corporate Leadership involve becoming a leader. Corporate Leadership envisions the commercial direction in which the company will evolve. Technical Fellows envision the technical direction in which the company will evolve

To lead, you must have a knowledge base broad enough to discern trends in national societies and national economies, then be able to integrate these trends to formulate a global picture of the future. Formulating such a picture requires an understanding of sociological and political trends. More importantly, it requires a broad understanding of scientific trends: which sciences and technologies are going to be economically controlling in the future. Thus, you function at the Macro level.

To be Technical Fellow requires developing leadership skills during your career, which means studying leaders and formulating a set of principles to govern your personal conduct. Reading biographies and autobiographies of acknowledged leaders will provide you examples of leadership and the skills required to be a leader. Such reading also provides you with "virtual" mentors.

A plethora of books on leadership exists --- which you need to read, particularly those entitled "the leadership principles of Churchill", "... of Washington", "... of Lincoln", "... of Grant", "... of Eisenhower", and so on.

To be a leader, you must formulate a set of core principles to govern your personal conduct and you must adhere to your principles. A leader is trusted by others; thus, to be a leader, you must develop a set of principles that will generate others' trust in you. All religions promulgate such a set of principles. In Christianity, that set of principles is the Ten Commandments. Reading philosophical ethics will also provide you material for developing a set of core principles. You cannot give "lip service" to your core principles: you cannot follow them on even days of the week and ignore them during the other days of the week. Others will trust you only if you adhere, faithfully, to your core principles.

Most Technical Fellows are recognized nationally as experts in a given process or product. Publishing refereed papers, technical articles, and/or patents provides such recognition. Presenting papers and leading discussions at professional society meetings, such as the AIChE and ACS, also provides national-level recognition.

Therefore, early in your career, you must

- formulate a set of core principles;
- read biographies and autobiographies of historically recognized leaders;
- read sociological and psychological literature;
- read macroeconomic literature;
- submit patent applications;
- write refereed papers and technical articles;
- participate in AIChE and/or ACS.

Career Development
Ten Year Goals --- Senior Process Engineer/Senior Research Chemist

Remember: at this level, the Senior Level, you identify trends that impact a business, process, or product. To identify trends that impact a business means interacting with a Business Group. Basically, this function identifies, justifies, prioritizes, and sells projects to a Business Group; i.e., gets a Business Group to authorize and fund a given project.

The best way to identify potential projects is to study the original DuPont Return on Investment (ROI) calculation, as shown in Figure 6. The ROI calculation shows the individual components controlling product costs. Potential projects exist in those cost numbers. Asking a manufacturing facility what equipment or process causes the most problems is another way to identify potential projects. Once a potential project has been identified, it must be justified; i.e., it must be analyzed economically. Such an analysis uses the tools you learned in your university Process Engineering course. After determining the economic value of a potential project, it is prioritized against existing potential projects. Net Present Value (NPV) is generally used for project prioritization. You will then present this prioritized list of projects to a Business Group biannually or annually. In other words, you will make a presentation to a Business Group that is actually a sales pitch. You will be selling projects to a Business Group. Such a presentation requires speaking skills. A project becomes “active” once a Business group authorizes funds for it. It then becomes your responsibility to manage the project and implement it commercially. Managing such a project means assembling and organizing a team of engineers and scientists.

To organize your project, use a project management software package, such as Microsoft Project. Outlining the work effort in Project shows the course and scope of the work effort and identifies the “rate limiting” steps embedded in the project, thereby allowing you to initiate their resolution before they become a crisis.

To assemble your project team, use concepts promulgated by Concurrent Engineering. If your project is for an existing manufacturing facility; i.e., process, include a plant process engineer on your team from its start. Doing so allows you to resolve plant issues as they arrive, rather than learning about them at the end of the effort, then restarting the project to resolve the identified issues. Including plant engineers in the initial project team accelerates development of the project and ensures and eases its implementation at an existing production facility. The same is true for Business Group personnel. Figure 7 shows a typical Process Development team.

Process development teams are generally large and decentralized with regard to location. Thus, keeping everybody informed requires effort on your part. Fortunately, e-mail exists today. Issue a weekly report to all concerned regarding progress per each item on your Microsoft Project timeline. Issuing a weekly report requires significant effort on your part, but it returns large dividends. First, everybody concerned can see the progress made weekly. Secondly, everybody can see who is becoming the “rate limiting” step and nobody likes such visibility; thus, everybody expends utmost effort on the project. Third, it is your the opportunity to write, thereby improving your ability to express yourself on paper or on an electronic monitor. Thus, early in your career, take as many writing courses as you can and read one or two how-to books on writing every year of your career.

Every project, at some time, requires face-to-face team meetings and teleconferences. With regard to such events, your goal is to keep the meeting or teleconference short and to the point. Invite only those necessary to make a decision and issue an agenda well ahead of convening the meeting. Identify on your agenda the issues requiring a decision. Ask participants to come ready to make the required decisions.

Therefore, to reach the Senior Level, you must learn to

- write quickly, clearly, and succinctly;
- make oral presentations;
- identify potential projects;
- justify potential projects;
- prioritize potential projects;
- sell potential projects;
- assemble multi-functional teams;
    - use Concurrent Engineering techniques;
- organize multi-functional teams;
    - use a project management software package;
- lead multi-functional teams.

You must also become an expert in one of the unit operations used in chemical processing and you must learn the scaling concepts for all the unit operations used in chemical processing.

Career Development
Five Year Goals

Your five year goals will be more detailed than your ten year goals. First, if you have not earned a Masters or Doctorate in Chemical Engineering, you should strongly consider earning a Masters or, at least, taking courses in Advanced Fluid Mechanics and Advanced Heat Transfer, as well as a course in Process Heat Transfer. These courses can be taken at a local engineering school where such courses are generally offered during evening hours. If you have a Doctorate in Physical Chemistry or Organic

Chemistry, then you should definitely consider earning a Masters in Chemical Engineering.

Second, if you are a plant engineer who wants to do Process Development, then you should inform management of your need for R&D experience. R&D experience will provide insight into the design of experiments necessary for Process Development. It will also provide insight about R&D's method for identifying, justifying, prioritizing, and executing projects. At the same time, you, as a plant engineer, will be able to educate the R&D staff about how plants view R&D projects, what concerns plants have about R&D projects, and what data and information plants require for approving and conducting plant tests, and how plants want the plant test organized, implemented, and executed.

If you are a process development R&D staff member without production plant experience, then you need to inform management that you desire a two-to-three year assignment at a manufacturing facility. Plant experience provides you insight into the issues that concern plant engineers about plant tests. Those issues are

- impact upon production;
- impact upon process safety;
- impact upon personnel safety;
- impact upon site environmental permit;
- equipment needs, temporary and permanent --- and the safe use thereof;
- data collection requirements;
- data analysis --- who analyzes the collected data;
- goals --- what constitutes success, who declares success or failure;
- commercialization of new process or altered process.

However, the main reason for plant engineers to get R&D experience and for R&D staff members to get manufacturing experience is to reduce the suspicion of each organization for the other. This suspicion exists because the mission of Technology Management is to wrought change while the mission of every manufacturing facility is not to change, is to continue producing product at maximum rate. By exchanging personnel, each organization gains insight into the purpose of the other organization.

Third, take the Professional Engineer's exam until you pass it. Passing it gives you *cache* with plant engineers and it provides you with more career opportunities in engineering. It also allows you to purchase liability insurance for your designs.

Fourth, learn and use a project management software program, such as Microsoft Project. Using such a program helps plan and organize a process development effort. It also identifies the "rate limiting" step of the project, thereby providing you a warning and allowing you to work the issue before it becomes a crisis. These programs also improve your productivity in general because they force you to plan and organize your work.

Fifth, hone your writing and presentation skills. At the Entry Level, you identify trends impacting individual processes or products. Generally, you will be asked to reduce the manufacturing cost of a process or product. Technical Resource Management will ask you to write progress reports as well as give presentations about you efforts to date. Use these opportunities to improve your writing and presentation skills. Read books and take courses of writing and presenting technical information.

Sixth, choose a unit operation or technical subject as you area of expertise. Your choice will depend upon which unit operations and technical subjects your employer uses, which might include some or all of the following:

- absorption;
- adsorption;
- corrosion;
- distillation;
- fluid dynamics;
- extraction;
- heat transfer;
- inspection/stress analysis;
- metallurgical analysis;
- process control/instrumentation;

- reaction engineering;
- thermodynamics.

Seventh, develop your Process Economics skills. If you did not take a Process Economics course at university, then enroll in an evening course at your local engineering school or take one of the short courses offered by various professional organizations. If you attended such a course during your undergraduate years, then consider taking an advanced course in Process Economics.

Eighth, during your early career years, you will participate on one or more multi-function teams. Observe how the team leader functions; analyze the leader's successes and failures and record your results in a "memo to file".

Ninth, join a professional organization, such as AIChE or ACS or both. Participate in the local organization in your area and in the national organization. These organizations provide you an outlet for technical articles as well as for making oral presentations. Becoming involved in their management provides you with an opportunity to develop leadership skills. Organize presentation sessions for their national conferences, chair such sessions, and join the leadership group of those subject areas which interest you. In particular, give attention to those sections covering Project Management.

Tenth, learn from your mistakes. Failed experiments teach us more than successful experiments do. Successful experiments simply confirm the validity of our theory and thinking. Failed experiments force us to reassess our assumptions and our theory. The same is true with respect to all endeavors. Endeavors evolving according to plan do not teach us much. Endeavors that go awry can educate you for future success --- if you learn from them. Thus, when an endeavor fails to meet expectation, take an afternoon to analyze what went wrong and why it went wrong. Be brutally honest with yourself. If you could have done better, then admit it. If you made errors in judgment, then admit it. Record your analysis in a recoverable form, either as an electronically stored document or as notes in a bound ledger; i.e., notebook. You will never improve until you admit that improvement is required. Your self-assessment provides you with data against which to measure future effort. Lastly, do not fear admitting mistakes to your supervisor. Freely admitting mistakes saves time and effort on everybody's part. It also shows that you assume responsibility for your actions --- a personality trait greatly admired by all supervisors.

Mentors and Mentees

Newly graduated engineers and scientists hear much today about the need for "finding" a mentor. They are told to find a mentor in order to reduce the time required to go from entry Level to Organization Leadership. However, mentees, those who are mentored, do not "find" mentors: mentors find mentees. Mentors are an organization's talent scouts. It is an unofficial role and a volunteer activity. A mentor is someone who keeps their eye open for the younger colleague demonstrating technical knowledge, personal responsibility, and self-initiated drive. Mentors know that such people require nurturing and promotion if the organization is to survive.

When newly graduated engineers and scientists hear that they need a mentor to accelerate their career, they assume the mentor has to be in Organization Leadership; i.e., has to be at least a Vice President; otherwise, how can a mentor accelerate their career to a similar lofty position. Unfortunately, newly graduated engineers and scientists rarely obtain the opportunity to present significant results; i.e., impress, Vice Presidents. Newly graduated engineers and scientists most often give presentations to Senior Level personnel. It is they whom newly graduated engineers and scientists impress. Thus, mentors are generally one level above a mentee. Senior Level personnel provide mentors for Entry Level personnel; Organization Leadership personnel provide mentors for Senior Level personnel.

Newly graduated engineers and scientists also assume that the mentor becomes responsible for their career development. Nothing could be further from the truth. Mentors provide promising individuals information about future technical and business trends, about organizational structure, and about how to navigate through that structure. Mentors do not take mentees by the hand and usher them into an Organization Leadership office. Mentors show mentees how to use information to produce the best possible work product.

Interestingly, while good mentors are "dime a dozen" --- good mentees are rare. The best documented career of a fully receptive, fully responding mentee is that of Dwight David (Ike) Eisenhower.[1] Ike was not the little known, lucky officer promoted to prominence as portrayed by many writers. Ike graduated from West Point in 1914 and was, by 1919, identified as a promising officer. In the early 1920's, Ike met Brigadier General Fox Connor, who became Ike's mentor. Figuratively, Connor poured information into an infinitely absorptive sponge, into Ike. And Ike responded by working hard to learn all that Connor could teach him. During the late 1920's, Connor passed Ike to the next level mentor: Army Chief of Staff Douglas MacArthur. MacArthur mentored Ike until mid-1939. Ike responded to MacArthur just as he had responded to Connor: he worked hard to learn all MacArthur could teach him. Throughout Ike's career, he was considered by those above him in rank to be the best, the outstanding officer of his age group and West Point class. He earned that reputation by working hard for his mentors, not by expecting those mentors to take his career in hand and walk him into a Leadership office. Within the Army, it was no surprise that Ike rose quickly through the ranks once war overtook America.

Leadership

Today, companies talk much about the need for leaders in order for commercial survival. Unfortunately, companies do little to train future leaders. At best, a company may issue a book about leadership, ask the recipients to read it, then schedule meetings to discuss the reading assignment. These meetings are generally led by someone who knows nothing about leadership. Thus, the exercise accomplishes little, except that the company can claim it has a leadership development program. The only organization that actively trains leaders from the day individuals join it is the American military. The American military understands the need for everybody to be a leader and it makes every attempt to provide leadership training.

Short of joining the military, being active in a professional organization provides an individual with the opportunity to learn and develop leadership skills. Such organizations provide opportunities to function at the Entry and Senior Levels outside a company. If leadership mistakes are made, as they generally are, they are made beyond company sight.

How do professional organizations provide leadership training? They provide leadership training by asking engineers and scientists to fill roles requiring others to accomplish given tasks --- without remuneration. Getting people to meet their goals without the "remuneration stick" requires leadership. Volunteers accomplish goals because they respect the person asking them to do so.

Professional organizations provide individuals the opportunity to learn what makes a leader. Thus, your participation in a professional organization benefits both you and your employer: you learn to be a leader; your employer gains a leader.

MBA: To Earn or Not to Earn

The question asked by every newly graduated engineer and scientist today is: "Should I earn an MBA?" Maybe it is not a question but rather a statement: "I plan to earn an MBA." During the early 1980's, few engineers or scientists had earned an MBA --- or even considered it necessary to do so. By the mid-1990's, most young engineers and scientists thought they needed an MBA to rise within a commercial organization. By the mid-2000's, almost all young engineers were pursuing an MBA in some kind of program: full-time, evenings, weekends, long-distance.

Three forces seem to be motivating young engineers and scientists to earn an MBA. First, realization during the 1990's that individuals on Wall Street and in corporate management were earning large salaries and exceptionally large year-end bonuses. Study of these individuals showed nearly all had an MBA. Lesson: if you want a chance at a hyper-inflated salary and bonus, then obtain an MBA. Second, from the end of WW II through the 1980's, when a young engineer or scientist looked up the corporate ladder, he or she saw each rung populated with Ph.D.'s. Thus, to move up that ladder, you considered earning a Ph.D. By the end of the 1990's, many rungs of the corporate ladder were populated by MBA's and by the end of the 2000's, almost all rungs were filled with MBA's. It is human nature to believe that if you have a Ph.D. it takes a Ph.D. to replace you. Any lesser degree devalues the effort you made to earn your Ph.D.

The same is true of an MBA: it takes an MBA to replace an MBA. If not, then why earn an MBA? Thus, the degrees populating a promotion ladder are self-propagating, self-fulfilling. This type of thinking is forcing all newly graduated engineers and scientists to consider earning an MBA once employed by a commercial organization. Third, the advent of the personal computer and spreadsheet software, such as Microsoft Excel, has made quantitative business calculations possible, as well as statistical analysis of large data sets. Much of an MBA program is teaching people to apply mathematics to business problems and to use the calculated results to formulate a decision. Prior to personal computers and spreadsheet software packages, such calculations were long, time consuming, and fraught with calculation errors. The result did not seem worth the effort, especially when applying statistical concepts to large data sets. Now, such calculations can be done in a fraction of a second --- so why not do them? In fact, today it is *de rigueur* to do them. Engineers and scientists are trained as undergraduates to use calculated results to make and support decisions. For them, applying the same thought processes to business-oriented problems is not a great step out of the proverbial conceptual box. However, such is not the case for art history, English literature, fine arts, history, political science, psychology, social work, sociology, or any number of other degrees that involve calculation-free thinking. For these latter degrees, earning an MBA is important because a new skill is learned --- calculation using spreadsheets. The degree of importance assigned to earning an MBA may not be as high for engineers and scientists since it does not necessarily provide them with an additional skill.

Conclusions

The purpose of this article is to help you plan, start, and develop a technical career in Process Development. The most important point of this article is: your career begins when you formulate a plan for its development. Prior to formulating such a plan, your personal development and career development are at the mercy of an array of forces dictating your response. With a career plan, you harness and direct those forces to shape your response to meet your personal development and career development needs.

As an engineer or scientist, your first choice when formulating a career plan is whether to remain an engineer or scientist or become a business person. This decision controls your career path to Organization Leadership in a commercial organization. Having made the choice to develop a technical career, you are immediately confronted with a second decision: to be in Technology Management or to move into Technical Resource Management.

Each of the above career paths will develop in response to the three basic levels of the private sector economy. Those levels are Process Economics, Microeconomics, and Macroeconomics. The Process Economics or Entry Level identifies trends impacting individual processes or products. This level requires a thorough understanding of engineering and scientific principles and their application to "real world" problems. In other words, this level stresses technical knowledge and technical competency. Your undergraduate education prepares you for this level. The Microeconomic or Senior Level identifies trends impacting business and process or product groups. Within a commercial organization, such as a corporation, it is a senior position. At this level, you are considered a technical expert in a given unit operation or engineering subject or scientific subject, such as catalyst chemistry, reaction mechanisms and kinetics, or thermodynamics. Also, at this level, you will be expected to identify, justify, prioritize, sell, and implement technical projects. Implementing projects means you need to know how to assemble a technical team and how to lead such a team. Therefore, to be qualified for the Senior Level, you must be able to

- write quickly, clearly, and succinctly;
- make oral presentations;
- identify potential projects;
- justify potential projects;
- prioritize potential projects;
- sell potential projects;
- assemble multi-functional teams;
    - use Concurrent Engineering techniques;
- organize multi-functional teams;

    - use a project management software package;
- lead multi-functional teams.

The highest level is the Macroeconomic or Organization Leadership. At this level, you identify trends impacting national economies and industrial groups. You must be able to discern trends in national societies and national economies at this level and integrate them into a global picture of the future. Formulating such a global picture requires an understanding of sociological and political trends and how such trends impact the direction the sciences and engineering technologies take. Such positions require a broad technical knowledge base, as well as a broad personal knowledge base. It also requires you to be a leader. Thus, you must develop leadership skills to function at this level of society. Thus, you must

- formulate a set of core principles;
- read biographies and autobiographies of historically recognized leaders;
- read sociological and psychological literature;
- read macroeconomic literature;
- submit patent applications;
- write refereed papers and technical articles;
- participate in AIChE and/or ACS.

Figure 1.
Generalized Organization of the Private Sector Economy.

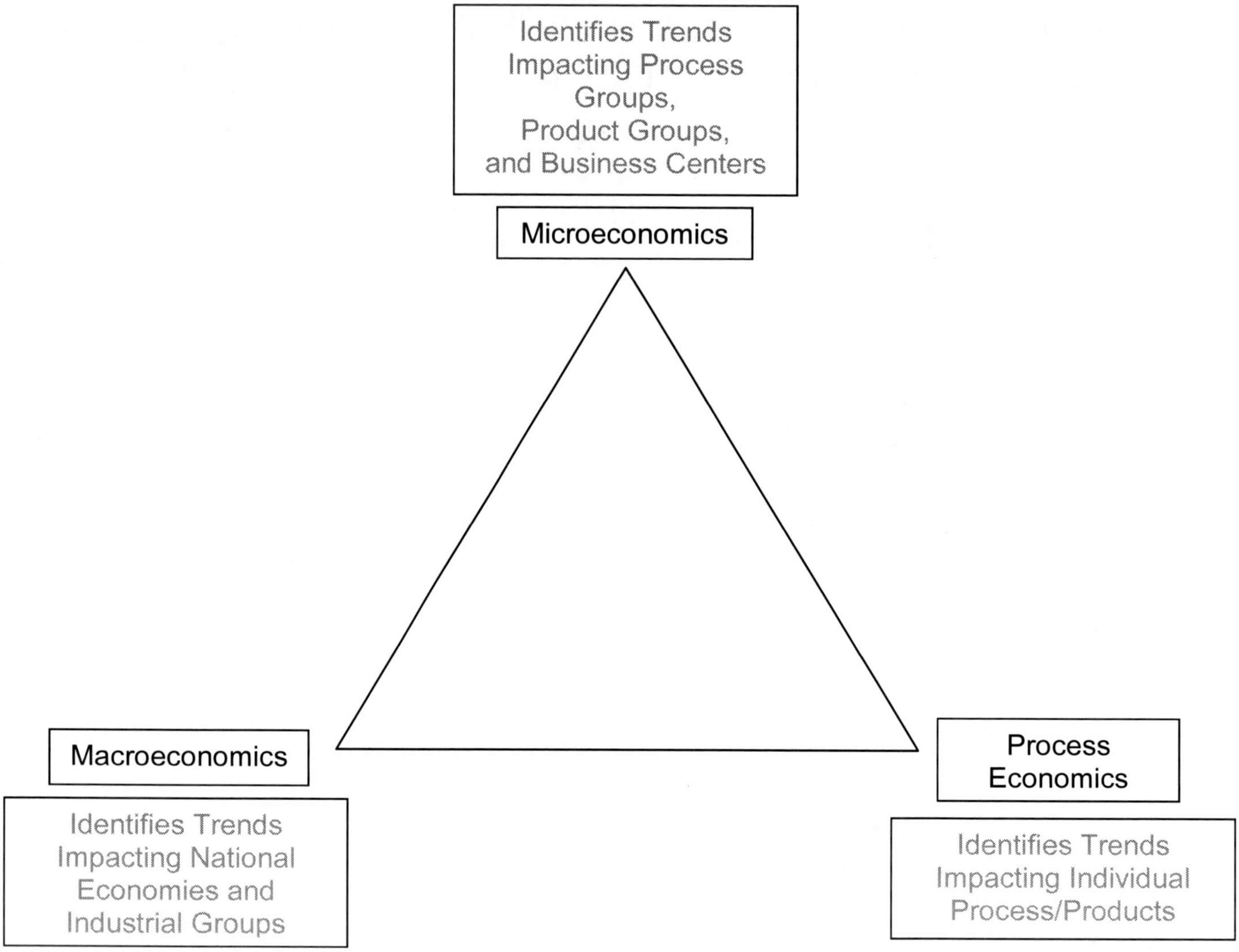

Figure 2.
General Structure of a Commercial Organization.

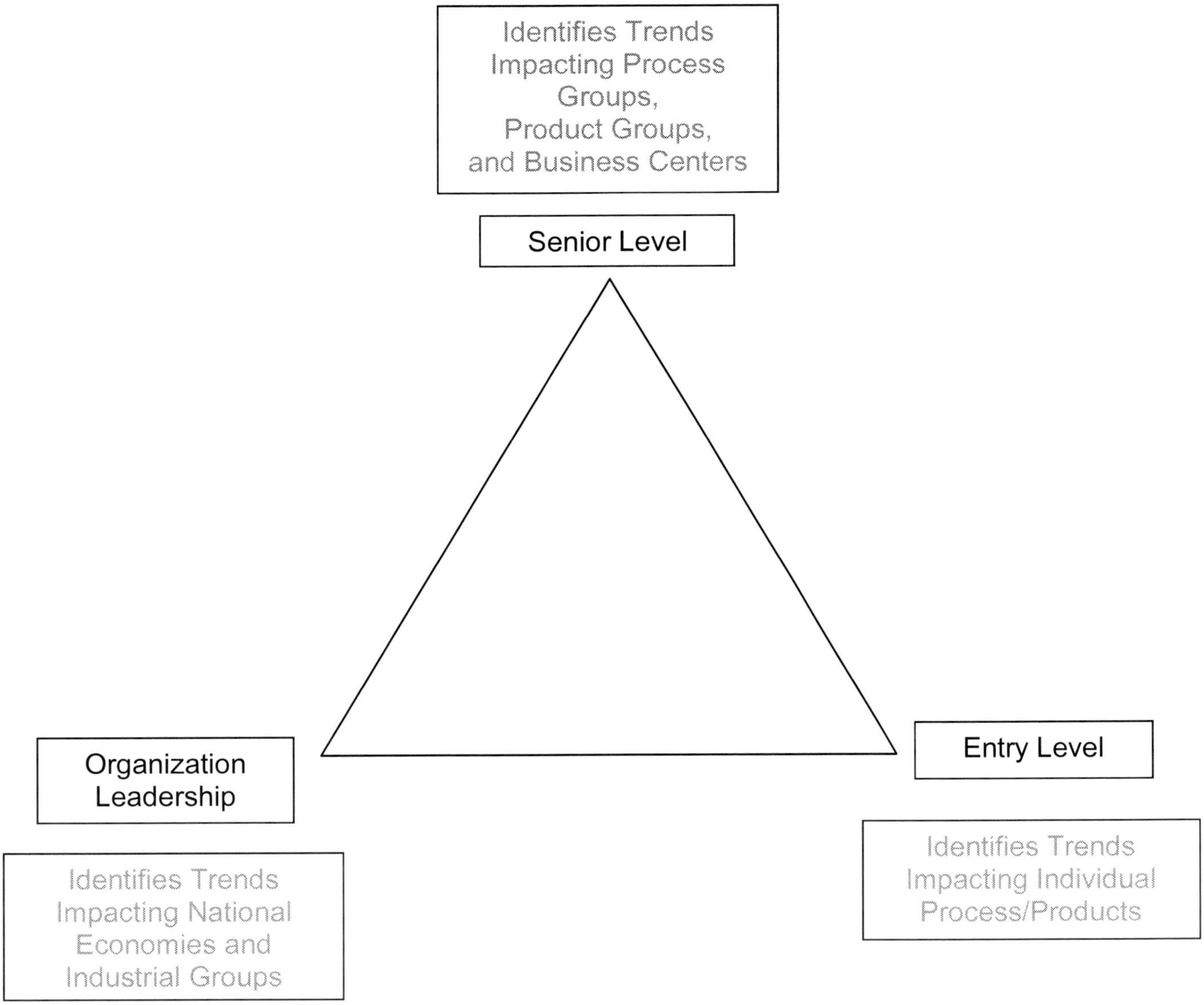

Figure 3.
Depiction of Various Career Paths.

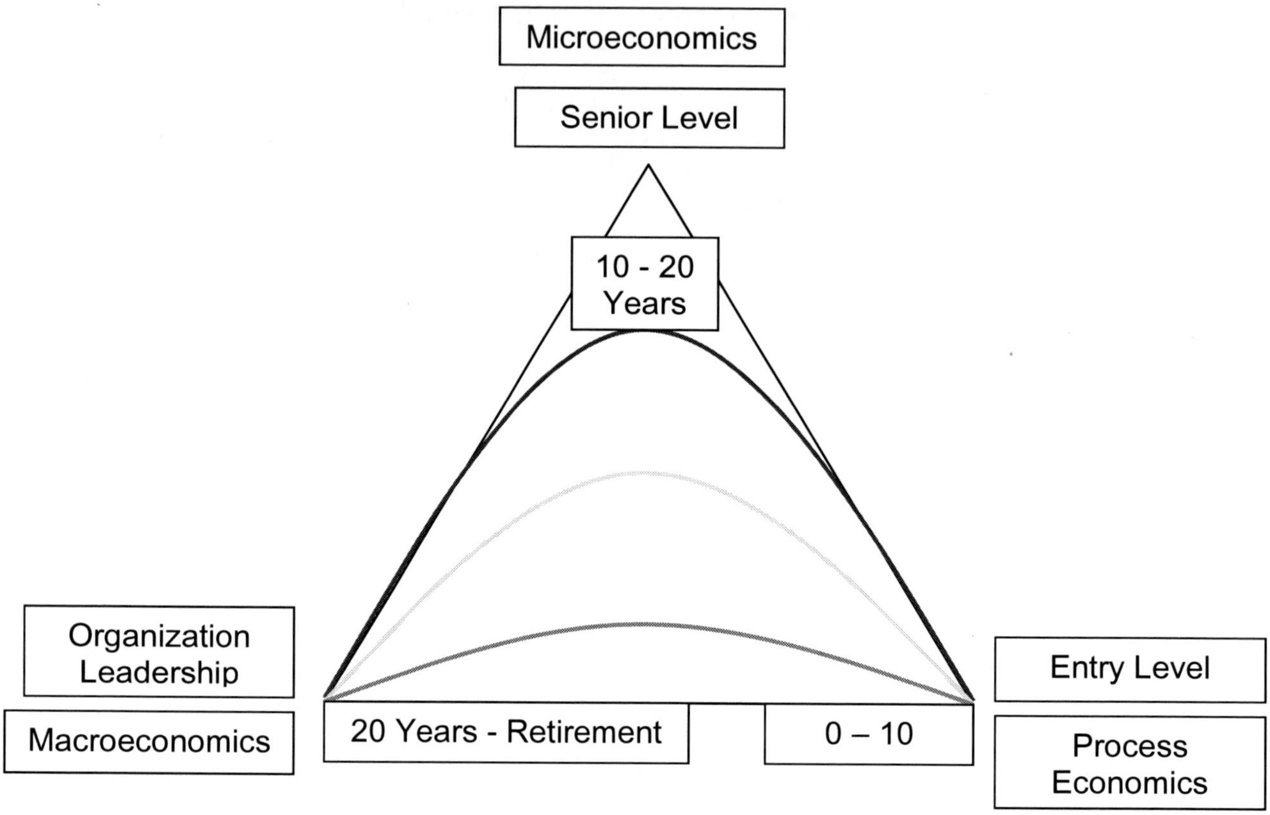

Figure 4.
Career Choices within Technical Companies.

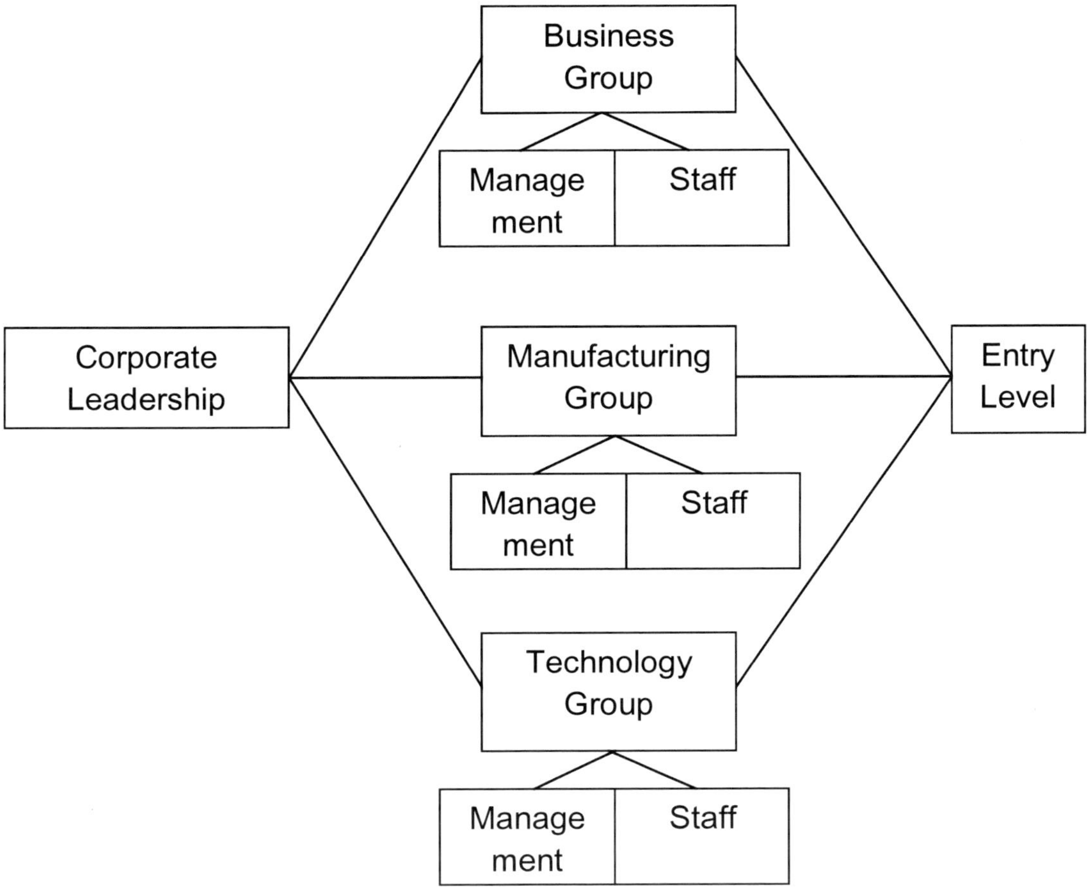

Figure 5.
Career Choices within Technical Companies.

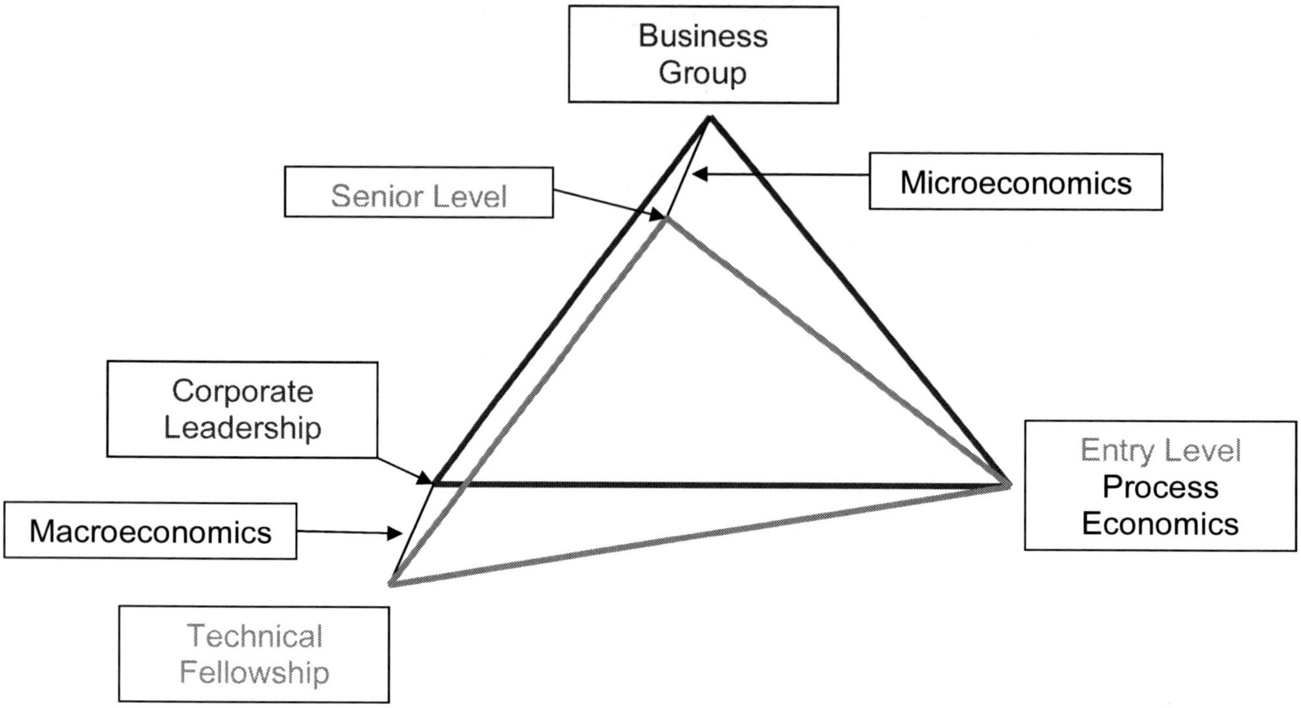

Figure 6.
Original DuPont Calculation of Return on Investment

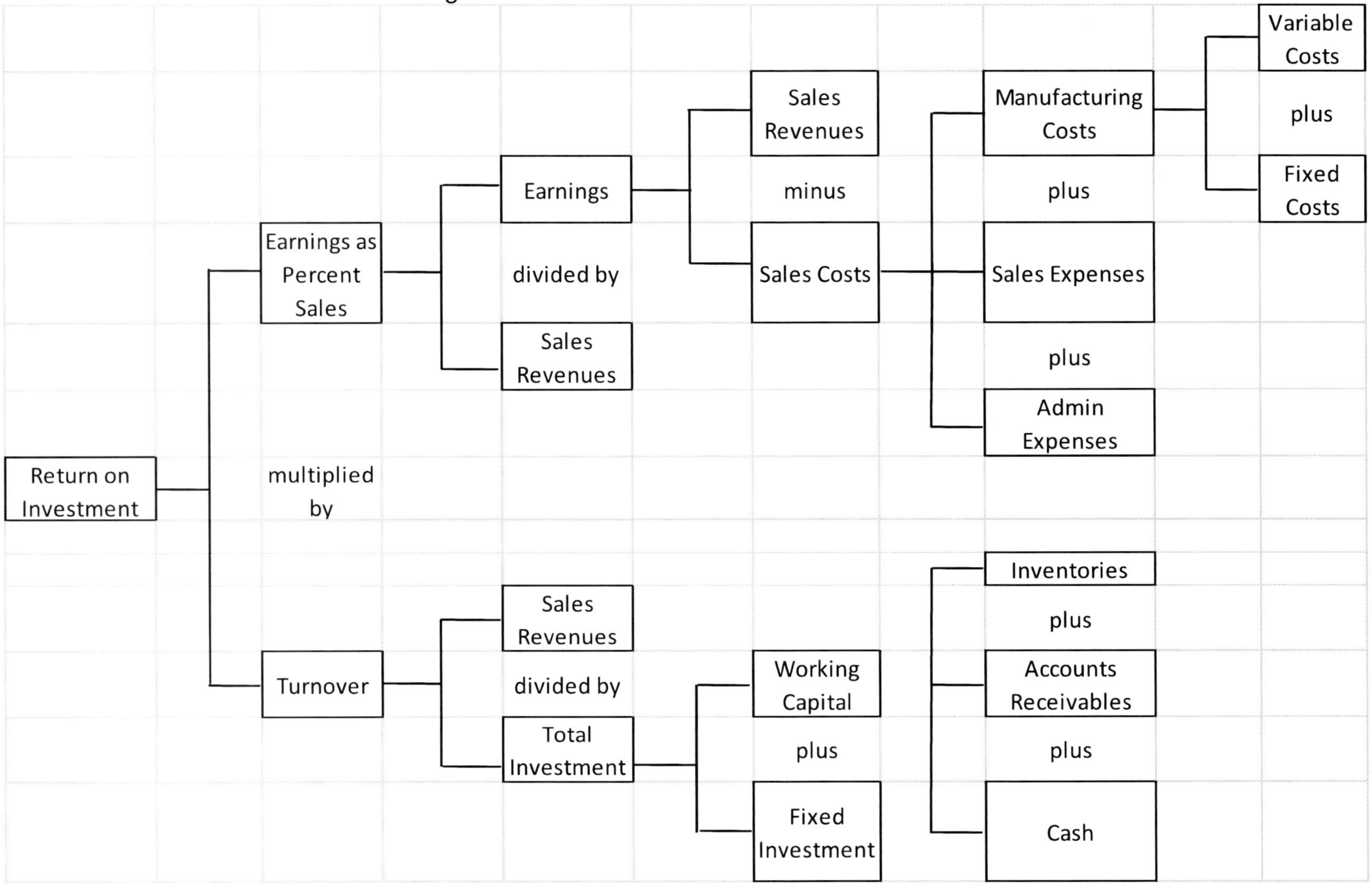

Figure 7.
A Typical Multi-functional Process Development Team

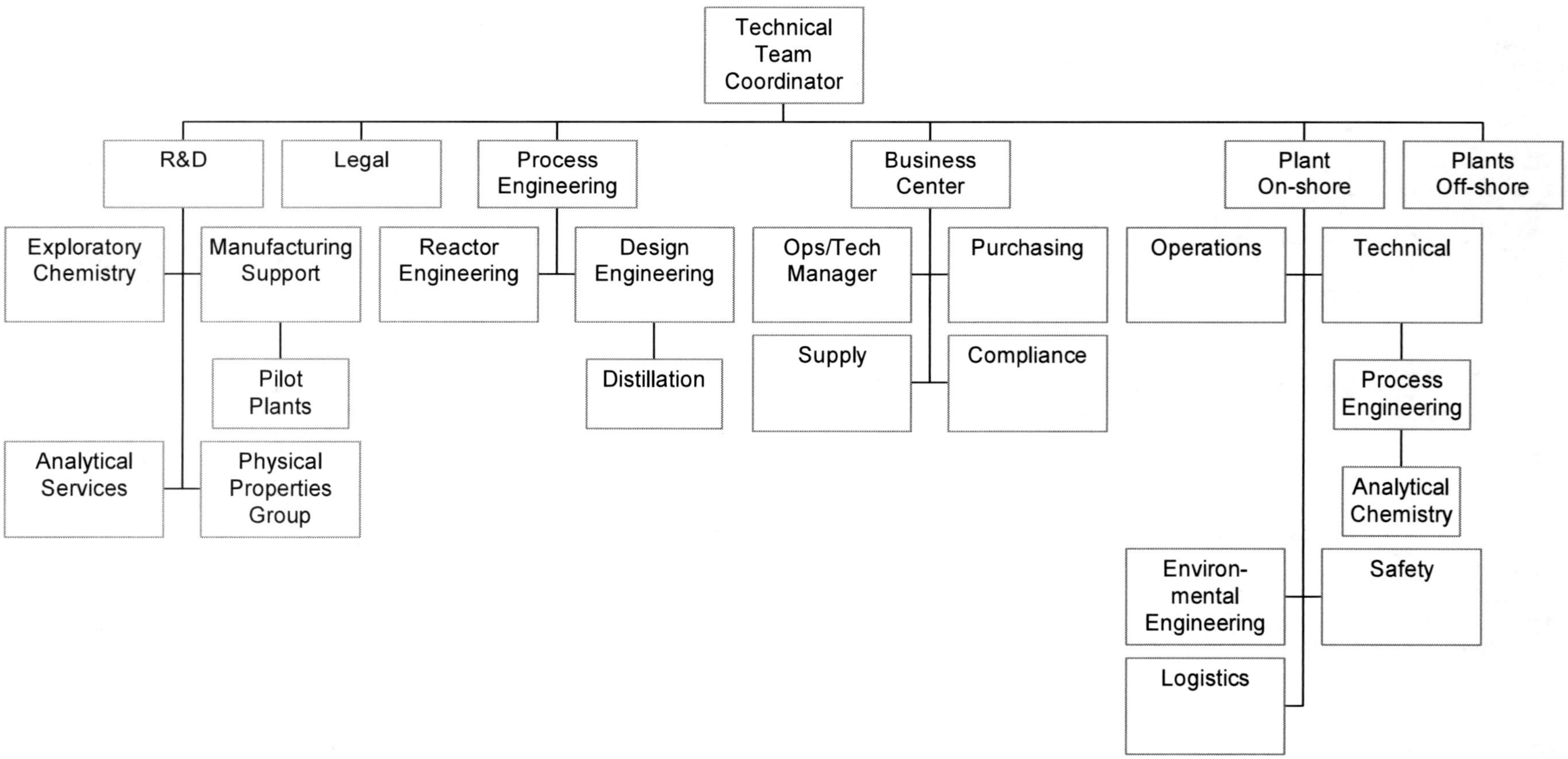

Process Development as an Industry Career Path

Joseph B. Powell, Ph.D.
Shell Chief Scientist – Chemical Engineering, Houston, TX

Introduction

While "process development" is not the subject of a separate course in the undergraduate or graduate-level chemical engineering curriculum, it plays a critical role in industry for the commercialization of new technology, and incremental improvement of existing technology. The following details activities, responsibilities, and opportunities associated with an industry career path in process development.

Process Development Discipline Areas

The relevant areas of process development were considered in setting up AIChE's Process Development Division 12 more than a decade ago, which grew from the "Pilot Plants" committee of the Engineering Group, with first chair dating back to 1984, to ultimately obtain full division status within AIChE in 1997. The initial areas commissioned for process development were [1]:

12A: Process Research and Innovation

12B: Pilot Plants

12C: Technology Transfer

12D: Manufacturing

While the division has grown to encompass Microprocess Engineering (12E), Process Intensification (12F), and Product Development (12G), the above four areas can be considered core to process development. Certainly, "pilot plants" is the most well-known sub-discipline under "process development". The construction and deployment of pilot facilities is the job description most commonly associated with this career path, and was indeed the founding basis for the process development division in AIChE. Scale-up and piloting of ideas generated from the "research department", for hand-off to "manufacturing", is one method by which process development can be implemented in industry. In companies where new product development is a core business opportunity, piloting of improved formulations can represent a full time opportunity. The various approaches to staffing and career paths for pilot plant support were documented in a benchmarking study by the Process Development Division's Area 12B [2], which identified however a broader definition of the role in many companies.

Stage Gates in Technology Development

There have been many names given to the various stages of technology development and deployment. Most recognizable for chemical engineering systems are:

- Research
- Development
- Demonstration
- Commercialization

Stages may be further defined in terms of readiness for deployment, including systems beyond those found in the chemical and process industries [3]. Processes developed by chemical engineers are typically sensitive to scale, and the above four categories are often used to depict the scale of study or implementation. Typically these steps entail bench scale research, one or more pilot plants (which are often the first integrated tests of process concept), followed by an optional demonstration unit to prove scale-ability or product viability, and finally, full commercial deployment. Rules of thumb for scale-up have been presented by Bisio and Kabel [4].

Process Development as the Integrating Discipline

While "development" is one of the stages along the path from idea to commercialization, it often plays an integrating role across all stages. Research may be driven for example by new chemistry, but in many cases the opportunity arises from a new catalyst, new process configuration, new unit operation such as a reactor or separation system design, or integration between unit operations as in process integration. Process development staff are therefore often highly involved in the research phase including idea generation, especially where engineering solutions are involved as part of the innovation.

"Development" leads to the initial prototype pilot units, but also an assessment of scale-up risk, from which the need for and sizing of demonstration units is decided. Core members of the process development team are therefore likely to be involved in demonstration projects, to reconcile scale-up behavior and process performance. New learnings also occur during commercialization [6], with six months typically required to achieve target rates and performance for a top quartile startup of new technology. The lead process development engineer and other core team members are also often involved with pre-commissioning, startup, and initial commercial operation, to effect "technology transfer" via extensive interaction with manufacturing.

Key chemical engineering sub-disciplines involved in these activities are the ability to do process synthesis for selection among competing options and unit operations, and economic analysis [5]. Chemical engineering training is perhaps unique among the engineering disciplines in emphasis on systems design and economics, and thus provides the basis for deciding between options, which can only occur after thorough grounding in business principles and drivers. Since all steps needed for commercial deployment comprise the process design, the process development engineer must become expert in virtually all aspects of chemical engineering, from chemistry and catalysis, through reactor design, optimization and selection among plausible separation options, as well as process control, metallurgy, safety and environmental stewardship. Interactions with demonstration or commercial project teams, engineering and construction partners, and ultimately a manufacturing location plus its business sponsors provide insight and training into the needs, activities, and work processes of the broader organization.

Along the complex path from idea to commercialization, a great number of skills are required, and typically gaps or "white spaces" arise to warrant learning of a new skill or core competency in fundamentals of chemical engineering, or other business, scientific, or management systems skills. Process development can thus be a visible springboard to a business career, a manufacturing location career, or a continuing career in R&D with focus on innovation, maintaining core technical skills, and knowledge of best practices in technology de-risking and development. Just as the AIChE Process Development Division has grown since 1984, opportunities in career development extend well beyond "pilot plants" for a career path in this discipline.

References

1. AIChE Process Development Division: http://www.pd-aiche.com/structure.html.
2. Bob Duggal (Afton), David Edwards and Sean Murray (Zeton), Daniel Pintar (UOP), Joseph Powell Shell), Pilot Plant Benchmarking Study Results. http://www.pd-aiche.com/area12b/pp_benchmarking_results.html
3. Mankins, John C. (6 April 1995). "*Technology Readiness Levels: A White Paper*". NASA, Office of Space Access and Technology, Advanced Concepts Office. http://www.hq.nasa.gov/office/codeq/trl/trl.pdf.
4. Bisio and R. L. Kabel, *Scaleup of Chemical Processes*, John Wiley & Sons, NY (1985).
5. J. B. Powell, *Economics of Pilot and Development Units for De-Risking of Commercial Processes,* **AIChE Spring National Meeting Conference Proceedings, San Antonio**, TX, *March 21-25, 2010.*
6. J. B. Powell and J. N. Chheda, *Assessing Potential Value of R&D Opportunities Via Explicit Assumption Vs. Implicit Experience- or Learning-Curve Approaches,* **AIChE Annual Meeting Conference Proceedings 2011**, October 16-21, Minneapolis Convention Center, Minneapolis, MN.

() Musings on Management

Frank van Lier, The Lubrizol Corporation, Wickliffe, OH

It really is all about the people!

Hiring good people.

Retaining them.

Adjusting for different personal styles, cultures, encouraging diversity, embracing different perspectives.

Motivating people to work together in successful global teams.

Achieving top notch results that are better than any individual can achieve alone.

Maintaining the focus.

Getting the results.

Management is all about the excitement managers gets from having their people and teams grow and succeed.

(82a) NREL's Integrated Biorefinery Research Facility - Pilot Scale Risk Mitigation

Jane C. Fisher, Daniel J. Schell, Richard T. Elander, David A. Sievers, Joseph Shekiro and Timothy Johnston, National Renewable Energy Laboratory, Golden, CO

NREL's Integrated Biorefinery Research Facility – Pilot Scale Risk Mitigation

Jane C. Fisher, P.E., Daniel J. Schell, Richard T. Elander, David A. Sievers, Joseph Shekiro, and Timothy Johnston, National Renewable Energy Laboratory

To achieve commercial-scale production of cellulosic ethanol at a cost that is competitive with gasoline, it is crucial to understand the entire integrated biorefining process and how one stage of the process can impact the performance of the others. With the addition of the Integrated Biorefinery Research Facility (IBRF) at the National Renewable Energy Laboratory (NREL), the cellulosic biofuels industry has access to a significantly expanded pilot plant and biochemical conversion process research facility. At NREL, the goal is to improve the cost effectiveness of cellulosic biofuels production processes and thereby to accelerate commercial scale deployment of these technologies. A number of aggressive government policies are guiding NREL's approach to cellulosic ethanol and other cellulosic biofuels research and development (R& D), including the 2007 Energy and Independence and Security Act requiring 36 billion gallons of renewable fuels by 2022, and President Obama's New Energy for America Plan calling for 60 billion gallons of advanced biofuels by 2030. The IBRF's R& D piloting capabilities will enable a wide variety of biofuels technology developers to reduce risks associated with scaling up to demonstration and full-commercial scales. The IBRF was completed in two stages and includes provisions to accommodate a broad and flexible range of equipment and processing options and configurations. These include front end capability to feed/mill a wide assortment of sustainable biomass feedstocks (corn stover, switch grass, sorghum, etc), horizontal and vertical reactor designs that enable a variety of biomass pretreatment operating conditions and chemistries to be evaluated, as well as the ability to perform high solids cellulose enzymatic hydrolysis. During completion of the first stage startup and commissioning a number of issues were encountered and to some extent resolved. Second stage design modifications were implemented to address issues experienced during the first stage. Design considerations as well as lessons learned during first and second stage facility commissioning, startup, and operation will be discussed.

Scaling Fixed-bed Catalyzed Processes

Jonathan H. Worstell

Consultant

Fixed-bed Reactors

Most common reactor for a continuous catalyzed process because

- Product is separated easily from catalyst;
- Capital requirement is lower than most other reactor types;
- Construction is easy;
- Sampling is easy;
- Plug flow can be assumed in most cases;
- Operation at high pressure is common.

Disadvantages of Fixed-bed Reactors

Large catalyst beds have poor heat transfer;

Channel flow can develop;

Product formation can be diffusion rate limited.

Scaling Fixed-bed Reactors

Hougen and Watson (1947)

Space Velocity

$m^3/hr/\ m^3 = 1/hr$

Same dimensionality as first-order reaction rate constant

Equivalence of Time

Space Velocity = Inverse Physical Time

Reaction Rate Constant = Inverse Chemical Time

Equivalent only if the process is
Reaction Rate Limited
First Order Reaction

Not equivalent if the process is diffusion rate limited

Catalysis Using a Porous Solid

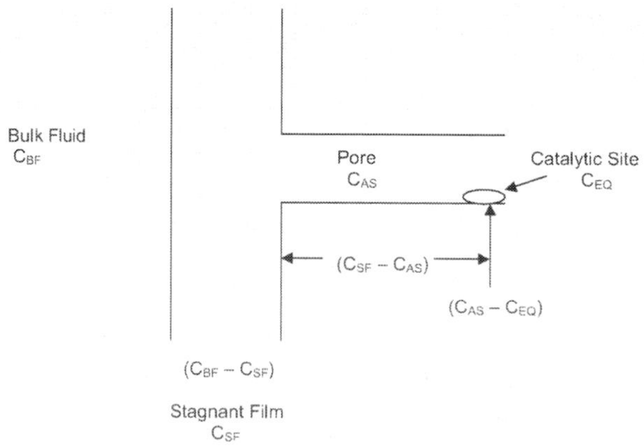

Boundary Layer Theory

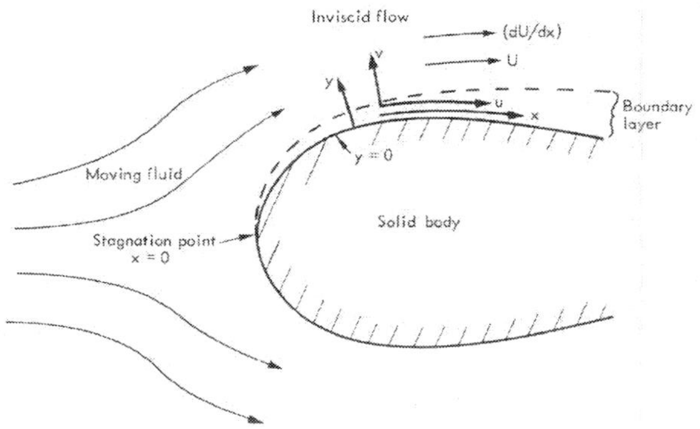

Various Rates for a Catalyzed Process Using a Porous Solid

$$R_{Rxn} = k_{Rxn}\left(C_{AS} - C_{EQ}\right)$$

$$R_{PD} = k_{PD}\left(\frac{a_P}{v_P}\right)\left(C_{SF} - C_{AS}\right)$$

$$R_{SF} = k_{SF}\left(\frac{S}{V}\right)\left(C_{BF} - C_{SF}\right)$$

Solving for the Known Concentrations

$$\frac{R_{Rxn}}{k_{Rxn}}+\frac{R_{PD}}{k_{PD}\left(a_P/v_P\right)}+\frac{R_{SF}}{k_{SF}\left(S/V\right)}=C_{BF}-C_{EQ}$$

$$R_{Rxn}=R_{PD}=R_{SF}$$

$$R\left\{\frac{1}{k_{Rxn}}+\frac{1}{k_{PD}\left(a_P/v_P\right)}+\frac{1}{k_{SF}\left(S/V\right)}\right\}=C_{BF}-C_{EQ}$$

Solving for the Global Rate Constant

$$R=\left\{\frac{1}{\frac{1}{k_{Rxn}}+\frac{1}{k_{PD}\left(a_P/v_P\right)}+\frac{1}{k_{SF}\left(S/V\right)}}\right\}\left(C_{BF}-C_{EQ}\right)$$

$$k_{Global}=\frac{1}{\frac{1}{k_{Rxn}}+\frac{1}{k_{PD}\left(a_P/v_P\right)}+\frac{1}{k_{SF}\left(S/V\right)}}$$

Simplifying the Global Rate Constant

$$\frac{1}{k_{Global}}=\frac{1}{k_{Rxn}}+\frac{1}{k_{PD}\left(\frac{a_P}{v_P}\right)}+\frac{1}{k_{SF}\left(\frac{S}{V}\right)}$$

$$\frac{1}{k_{Global}}=\frac{k_{PD}\left(\frac{a_P}{v_P}\right)+k_{Rxn}}{k_{PD}\left(\frac{a_P}{v_P}\right)k_{Rxn}}+\frac{1}{k_{SF}\left(\frac{S}{V}\right)}$$

$$\frac{1}{k_{Global}}=\frac{1}{\eta k_{Rxn}}+\frac{1}{k_{SF}\left(\frac{S}{V}\right)}$$

Relation of Various Rates

$$\frac{1}{\eta k_{Rxn}}+\frac{1}{k_{SF}\left(\frac{S}{V}\right)} \rangle \frac{1}{k_{Rxn}}$$

$$k_{Global}C_{BF} \langle k_{Rxn}C_{BF}$$

$$R_{Diffusion} \langle R_{Reaction}$$

If diffusion is present, can not use .
Space Velocity to scale fixed-bed reactors

Comparison of Space Velocities

Reactor Scale	Space Velocity (1/hr)	Linear Fluid Velocity (cm/hr)
Commercial --- Plant 1	18	16,500
Commercial --- Plant 2	11	9443
Recycle Pilot Plant	13	423
Catalyst Test Unit Pilot Plant	10	40
Laboratory	60	731
Laboratory	10	219

Concept of Boundary Layer

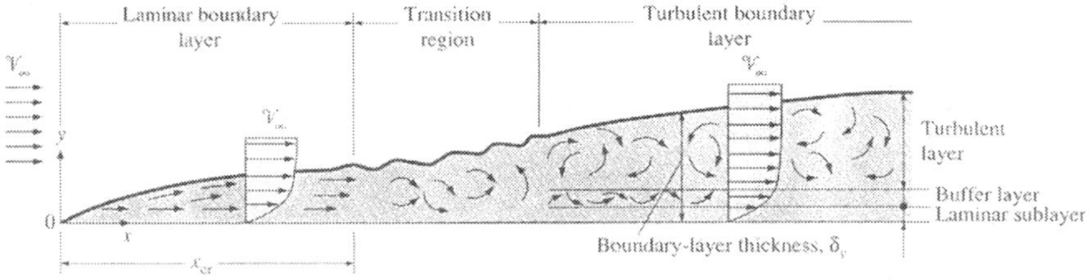

New Fixed-bed Scaling Parameter

$$j_D = \frac{k_{SF}\rho}{G}\left(\frac{S_{Eff}/V}{S_{Total}/V}\right)\left(\frac{\mu}{\rho D}\right)^{0.666}$$

$$j_D = \frac{0.455}{\varepsilon}\left(\frac{d_P G}{\mu}\right)^{-0.407}$$

$$k_{SF}\left(\frac{S_{Eff}}{V}\right) = 0.455\left(\frac{G}{\varepsilon\rho}\right)\left(\frac{S_{Total}}{V}\right)\left(\frac{\mu}{\rho D}\right)^{-0.666}\left(\frac{d_P G}{\mu}\right)^{-0.407}$$

from: J.M. Smith, *Chemical Engineering Kinetics*, 3rd Edn, McGraw-Hill Book Co, 1983.

Global Rate Constant vs Linear Fluid Velocity

$$k_{SF}\left(\frac{S_{Eff}}{V}\right) = \frac{v}{\beta}$$

$$\frac{1}{\beta} = 0.455\left(\frac{S_{Total}}{V}\right)\left(\frac{\mu}{\rho D}\right)^{-0.666}\left(\frac{d_P G}{\mu}\right)^{-0.407}$$

$$\frac{1}{k_{Global}} = \frac{1}{\eta k_{Rxn}} + \frac{\beta}{v}$$

Figure 2.

Inverse Global Rate Constant as a Function of Inverse Linear Fluid Velocity

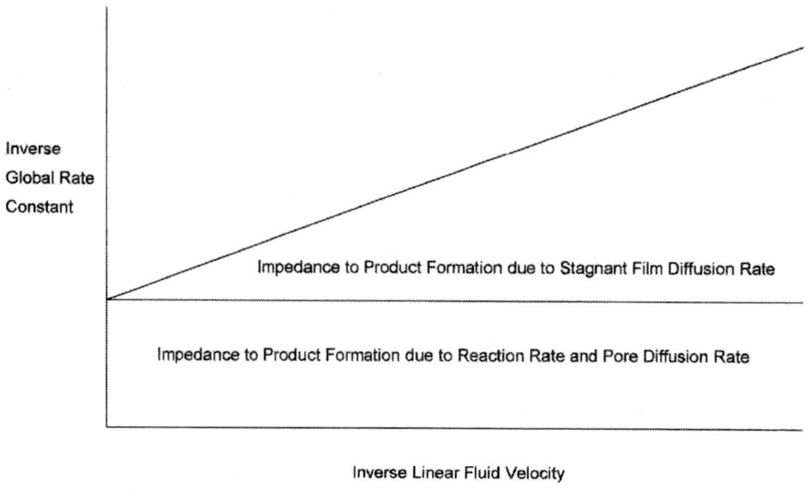

Most Common Design Flaw of Fixed-bed Reactors

Not recognizing that product formation is diffusion rate limited.

Definitely a problem when downscaling a fixed-bed reactor.

Scaling Fixed-bed Reactors
Summary

- Determine 1/k(Global) as a function of 1/v;
- Develop regression between 1/k(Global) and 1/v;
- Calculate 1/k(Global) at the proposed 1/v for the pilot plant (for downscaling) and for the commercial plant (for upscaling);
- Substitute k(Global) into the computer simulation to determine the size of the catalyst charge to the reactor;

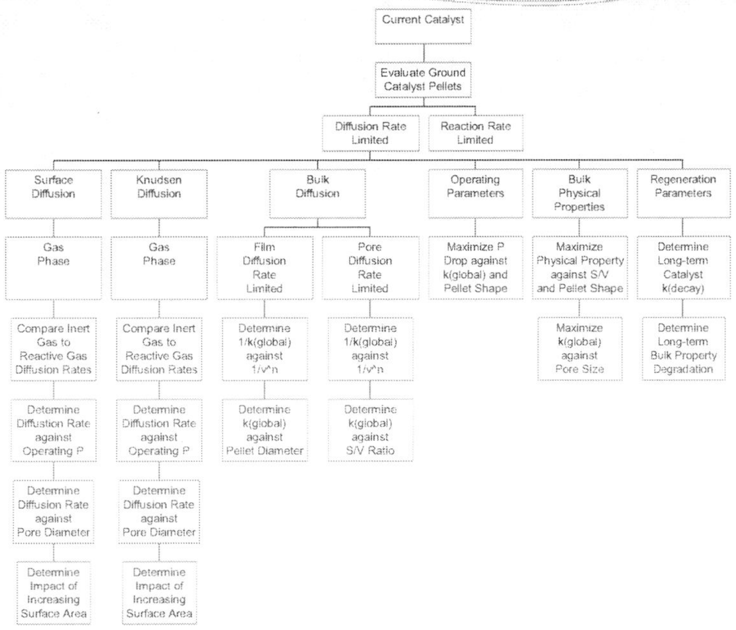

Scaling-up Fixed-bed Reactors

- Determine the reaction rate constant, k(rate)
- Determine the catalyst decay rate constant, k(decay)
- Develop a computer simulation using k(rate), k(decay), and volumetric flow rate to size the catalyst bed

Determining k(rate) and k(decay)

via Levenspiel's Method

Assume first-order decay of catalyst activity

$$\frac{dA}{dt} = -k(decay)A$$

$$\ln\left(\frac{A}{A_0}\right) = -k(decay)t$$

$$A = A_0 e^{-k(decay)t}$$

Determining k(rate) and k(decay) (con'd)

Assume first-order kinetics for reactant consumption

$$\frac{dC}{dt} = -k(Global)AC$$

Mass Balance for a fixed-bed reactor

$$\frac{W}{F} = \int \frac{dC}{-k(Global)AC}$$

Determining k(rate) and k(decay) (con'd)

$$-\frac{Wk(Global)}{F}\left(A_0 e^{-k(decay)t}\right) = \int \frac{dC}{C}$$

$$-\frac{Wk(Global)A_0}{F}\left(e^{-k_D(decay)t}\right) = \ln\left(\frac{C}{C_0}\right)$$

$$\ln\left\{\frac{Wk(Global)A_0}{F}\right\} - k_D(decay)t = \ln\ln\left(\frac{C_0}{C}\right)$$

Designing the Commercial Scale Fixed-bed Reactor

Determine the catalyst extrudate shape as a function of allowable pressure drop

Determine the required bulk physical property as a function of surface area to volume ratio

Allowable Pressure Drop as a Function of k(rate)

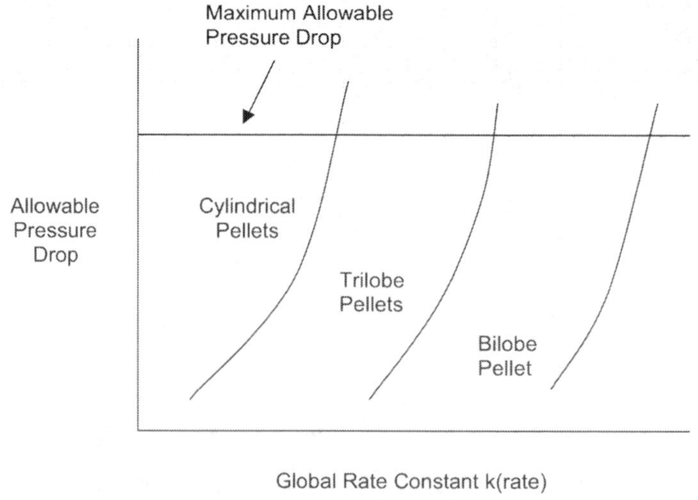

Bulk Physical Properties as a Function of Extrudate Surface Area to Volume

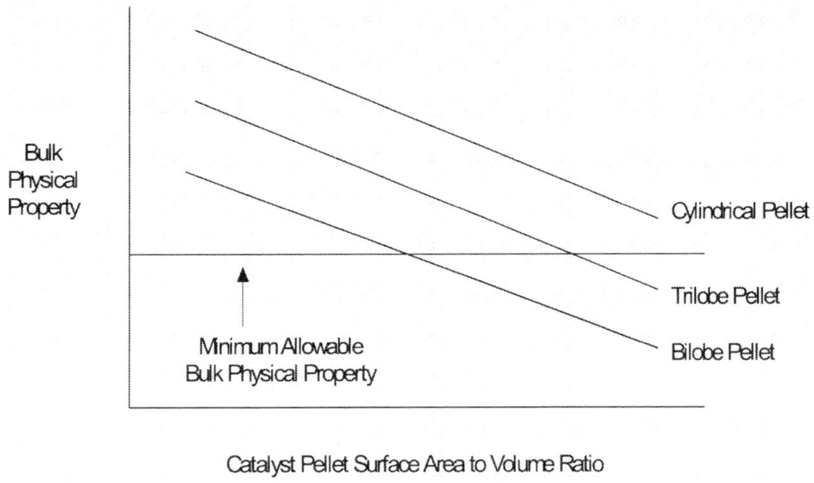

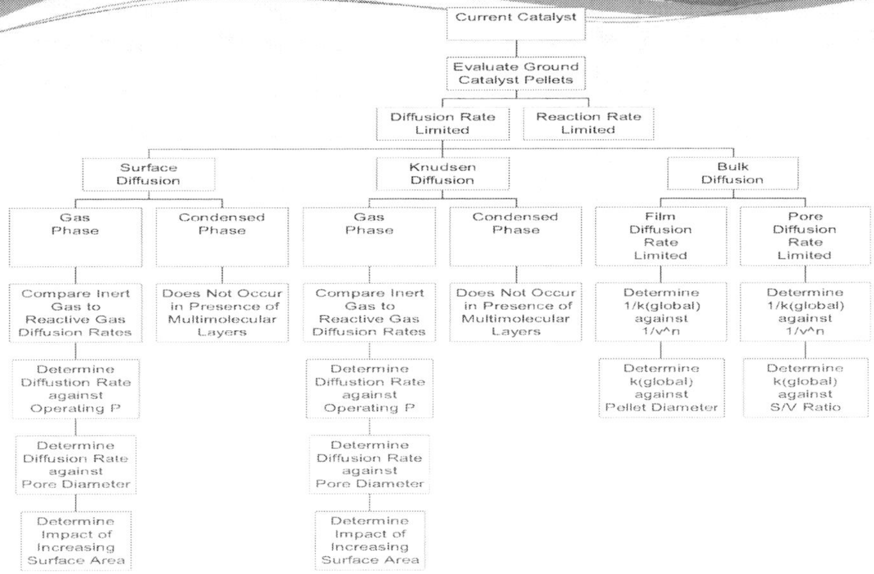

(82c) Computational Fluid Dynamics Modeling of Different Structured Reactor Packing for Process Intensification

Jeff Harris, Southwest Research Institute, San Antonio, TX, **Maoqi Feng**, Div. of Chemistry and Chemical Engineering, Southwest Research Institute, San Antonio, TX and Reggie Zhan, Emissions and Vehicle Research Division, Southwest Research Institute, San Antonio, TX

In this paper, Computational fluid dynamics (CFD)has been used to model temperature distribution, gas flow behavior, and pressure drop within three configurations of structured fixed-bed reactor packing: rod-shape packing with multiple channels (A), cake-shape packing (B) with multiple channels, and ceramic foam (C). It was found that all three configurations showed similar phenomena of the vortices on the outer edge just before the packing material. This is due to the flow encountering a sudden resistance causing the flow to change directions and swirl causing an increase in mixing of the two gasses. Although these vortices increased the mixing of the gasses, for Configuration A, the fluids would no longer be able to mix after entering the small channels because they were segregated from the rest of the flow. Configuration B was analyzed under the premise that each inserted disk would create a set of vortices to mix the fluid but would not segregate each flow through the rest of the length. The fluid will mix between each inserted disk and more disks can be added to further increase the mixing until a uniform concentration is achieved. Pressure drop determination by experimental validation was obtained.

A small scale pilot plant demonstration – case study of the synthetisation of 2-Ethoxy-2-methylbutane

Aarne T. Sundberg*, Aalto University, Finland
Petri Uusi-Kyyny, Aalto University, Finland
Ville Alopaeus, Aalto University, Finland

Abstract

A complete small scale pilot plant involving a recycle flow was assembled and tested using the synthesis of 2-Ethoxy-2-methylbutane as a demonstration case. Main components of the the pilot plant were a tubular reactor and a distillation column, accompanied with all the required auxiliary equipment. The distillation column was planar in shape and it was operated in a horizontal orientation. The column was packed with porous metal foam. The liquid hold-up of the system, excluding the feed and the product containers, was less than 50 cm^3. The concentration of the pilot plant was monitored using two online gas chromatographies. Stability of the miniplant during 50-hour long experimental runs was excellent.

Introduction

When developing a new or modifying an existing process a number of issues must be clarified before scaling-up the process into the commercial scale. A system which produces the desired product with a good yield in the laboratory scale might be difficult or unsafe to operate in the larger scale. As the size of the reactor is increased, the surface-to-volume ratio is decreased. Together with the less efficient mixing these may lead to runaway reaction. In addition to these challenges the accumulation of impurities caused by the use of industrial grade feed or by side reactions may cause unexpected problems. To detect these phenomena as early as possible the process is typically tested in one or more different scale pilot plants. [1-3].

The delay before the problems related to the accumulation of impurities can be detected may be up to several months. However, the separation efficiency of the distillation column may begin to deteriorate much earlier, but the reason might remain unknown. Lengthy pilot plant experiments could be required in order to detect such problems. [5]. The cost of these experiments can be quite high if performed on a large pilot plant. Especially, as due to the safety issues constant supervision of the process by well trained staff is required. To reduce these costs we have developed a small scale pilot plant, consisting of a reactor and a distillation column with reflux and boilup circulations and flow ratio controls. It was operated in a stable manner for over 50 hours. The liquid holdup of the system was below 50 cm^3, excluding feed and product tanks. Thanks to the low liquid holdup and the stable operation of the distillation column, the monitoring requirements could be relaxed. Thus the pilot plant was kept running overnight and over the weekends with no supervision.

Materials and methods

The process diagram of the distillation system is presented in **Figure 1**. The feed was pumped using a syringe pump (ISCO 500D) with a volume of 500 cm^3. For practical reasons, feed flow rates were selected to allow constant flow overnight without the need of refilling. Feed flow rate was controlled from the feed pump. The feed was preheated to the bubble point temperature of the mixture before entering the column.

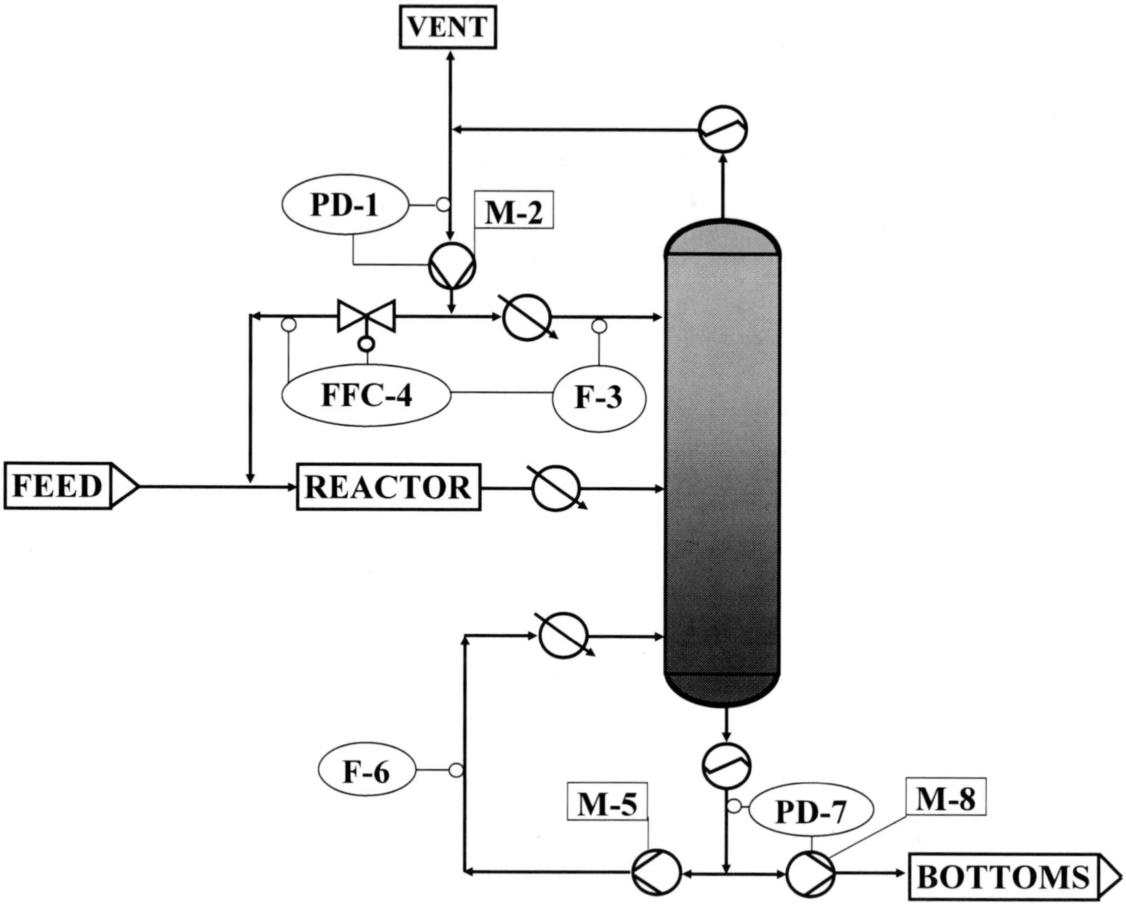

Figure 1. Distillation column process diagram.

All product flows were condensed and cooled to room temperature to protect pumps and instrumentation. Non-condensing gases were vented after the condensator. Liquid level inside the condensate pipeline was measured with a differential pressure transmitter, **PD-1,** (Huba Control, 692-series). The transmitter was coupled with the condensate pump, **M-2,** (HNP Mikrosysteme) and the liquid level measurement was used to control the rotational speed of the pump. The condensate flow was divided into reflux and distillate flows using two mass flow controllers. The distillate flow was either taken out of the system or mixed with the feed flow before the reactor. The reflux mass flow controller, **F-3,** (Bronkhorst, MiniCoriflow) was directly coupled with the distillate mass flow controller, **FFC-4**. Thanks to the coupling, the distillate flow controller could be operated either in a constant flow or as a flow ratio controller.

The heavy fraction was taken out from the bottom of the column by gravity. The reboiler pump, **M-5,** was operated with a constant setting with the mass flow controller (**F-6**) set to measure the flow rate. Liquid level in the reboil pipeline was measured with a differential pressure transmitter, **PD-7**. The transmitter was coupled with the bottom product pump, **M-8**, and the liquid level measurement was used to control the rotational speed of the pump. Thus, the bottom product flow was used as an overflow out of the system.

Three-way valves were used to manually toggle the concentration measurement between the reboiler flow and the reactor product flows. The concentration of the reflux flow was measured continuously. The tubing was made out of OD 1/8" Teflon®PFA, with the exception of connections between the heaters and the distillation column. All connections were made using compression fittings (Swagelok).

The distillation column was planar in shape, as presented in **Figure 2**. It was similar to the one presented in our earlier work, with some modifications [4]. The column was constructed by welding from thin sheets of stainless steel. The outside measurements were 300 mm by 50 mm by 20 mm and the chamber inside the column was 290 mm by 30 mm by 5 mm. Total volume of the column was 43.5 cm^3. The column was packed with a 5-mm-thick layer of open-cell, porous metal foam (Recemat RCM-NC-2733.03). The column had 12 pipe connections on the side for liquid inlets and for temperature measurements. The five connections on the top were for sampling and for vapour inlet and outlet. The single connection on the bottom, at the hot end, was for liquid outlet. The column was insulated using a 2-cm-thick layer of rock wool. Heat losses were further reduced with two heated aluminum blocks, placed above and below the insulation, as presented in **Figure 3**.

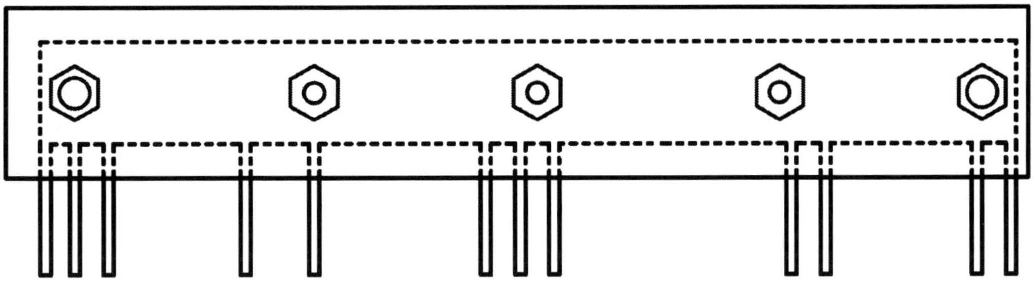

Figure 2. Diagram of the distillation column.

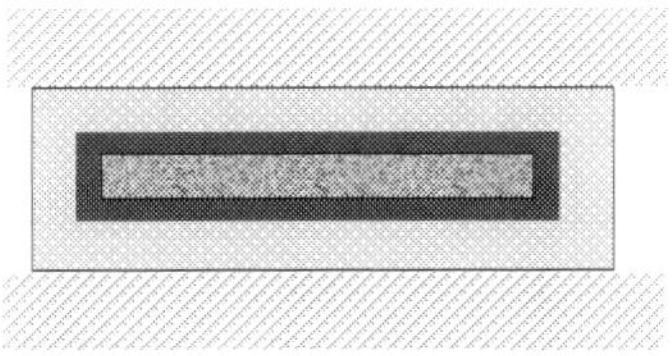

Figure 3. Cutout of the distillation column with insulation and heated aluminum blocks.

Chemicals. Ethanol (Altia) purity was over 99.5 mass-% and it was dried over molecular sieves (Merck, 3A) before use. Amberlyst 16 catalyst (Rohm and Haas) was first rinsed with distilled water, then with ethanol. Finally, it was kept overnight in ethanol. 2-methylbut-2-ene (Aldrich) was over 95 mass-% and it was used as supplied.

Analysis. Concentration was measured online using two gas chromatographies (HP 6950 and Agilent 6850) at three points; reactor product, distillate and bottom product. Additionally, when the system had reached steady state the concentration profile of the distillation column was measured from samples taken manually at 3 points along the column.

Preheaters and heat exchangers were manufactured by a milling machine (High-Z S-720) from a block of brass. Outer dimensions were 110 by 20 by 50 mm^3. The process fluid flowed through a tortuous channel with a length of 0.4 m. The depth of the channel was 0.3 mm for the heater, and 0.7 mm for the heat exchanger. Total volume of the process fluid was estimated to be less than 0.5 cm^3 for the heater and less than 1.0 cm^3 for the heat exchanger. The top and bottom plates were soldered over the channels.

Reactor. The reactor was constructed from a 12-mm-thick stainless steel tube with a wall thickness of 1 mm. The length of the reactor was 270 mm. The ends were constructed by a compression fitting type reducer with a filter inside the downstream end (Swagelok). The amount of catalyst used in the experiments was 12 g. The feed was preheated inside a 2-meter long loop of 1/16" OD steel tube. Both the preheating loop and the reactor were immersed in a heated water bath. The temperature of the water bath was kept at a constant temperature (+70 °C) in all of the experiments.

Instrumentation. Temperature inside the distillation column was measured at 8 points using NiCr-Ni thermoelements (Nokeval) and logged with a control system (Labview). The heating and the evaporating devices were controlled with the same control system.

Results and discussion

The feed was an equimolar mixture of Ethanol and 2-Methylbut-2-ene. The feed flow rate was set to 0.2 cm^3/min (0.14 g/min, 2.4 mmol/min). Reactor temperature was kept at +70 °C. The pressure of the reactor, which was monitored from the feed pump pressure meter, was about 3 bar. The composition of the reactor product was monitored at the beginning of the experiment and on the following day, as presented in **Figure 4**. Average 2-Ethoxy-2-methylbutane concentration was 36.8 mol-%. Conversion of the reactants was 36.8%, with close to 100% yield. The amount of impurities was below 0.2 mol-%.

In this work both the distillate and the bottom product were removed from the system. Both product flows were collected and weighted. The weighted results were compared with the counter of mass flow controller. Weighted results compared well for the bottom product, with an error of less than 1%. However, the distillate mass flow controller counter was consistently significantly higher than the weighted value, by up to

40%. This was probably due to the low bubble point temperature of the distillate, which caused the liquid to evaporate inside the mass flow controller, and the pressure fluctuations caused by the pump. The overall mass balance of the system was accurate to 99.8%, based on weighted results.

Concentration of the distillate and the bottom product is presented in **Figure 5 and 6**. The stability of the distillation column was excellent, as seen especially in the distillate concentration. The bottom product had a slight disturbance at 50 hours, caused by pumping problems. Time interval between measurements was about 15 minutes and more than 300 data points are presented. The time to reach stable operation after a disturbance was at most 3-4 hours. The system was kept running over 50 hours. After the initial phase, the operation was extremely stable.

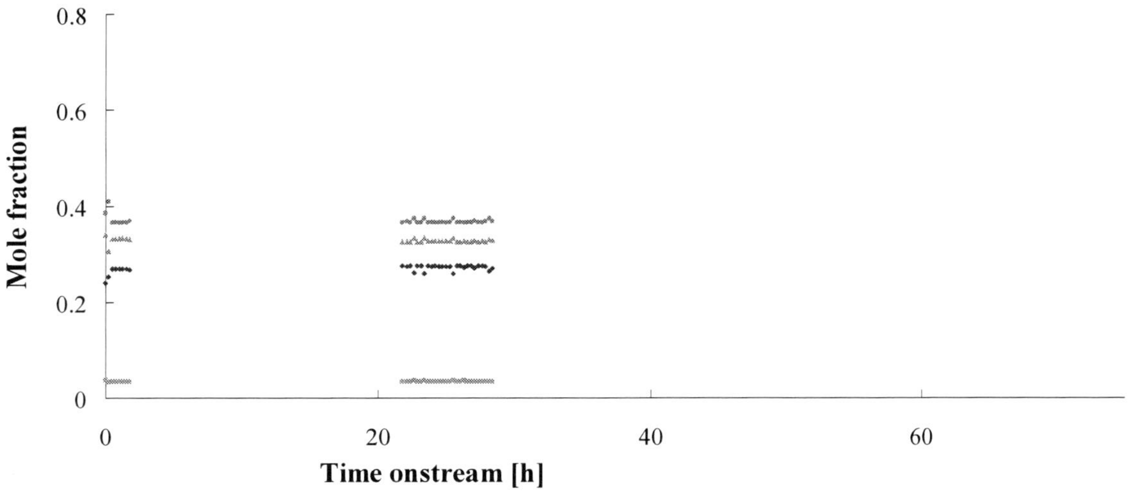

Figure 4. Composition of reactor produce flow. 2-Methylbut-2-ene, ▲; Ethanol, ◆; 2-Methylbut-1-ene, ■; 2-Ethoxy-2-methylbutane, ●.

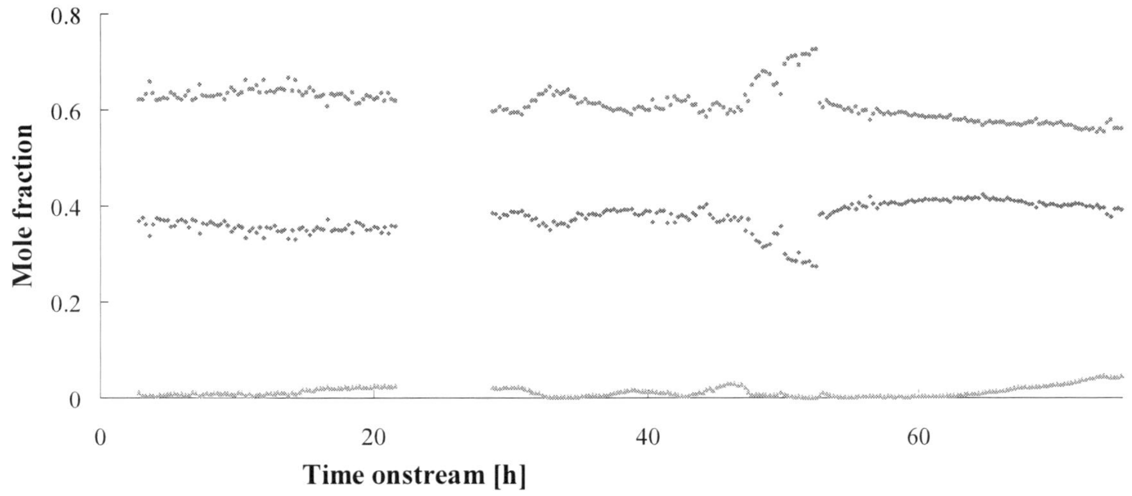

Figure 5. Composition of bottom product. 2-Methylbut-2-ene, ▲; Ethanol, ◆; 2-Ethoxy-2-methylbutane, ●.

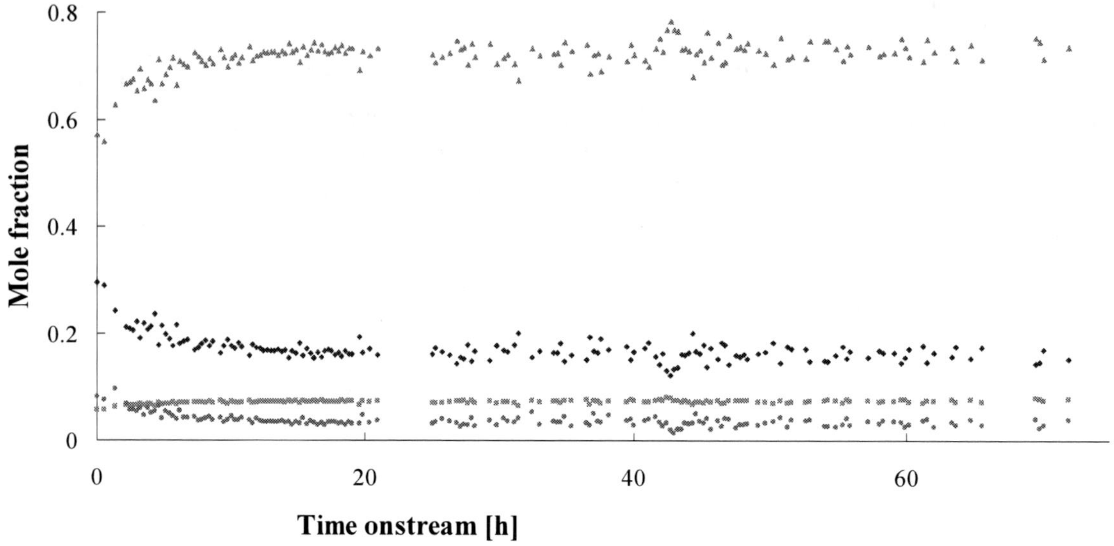

Figure 6. Composition of distillate. 2-Methylbut-2-ene, ▲; Ethanol, ◆; 2-Methylbut-1-ene, ■; 2-Ethoxy-2-methylbutane, ●.

Conclusions

A very small scale pilot plant was designed, assembled and tested. The liquid holdup of the system was below 50 cm^3. Synthesis of 2-Ethoxy-2-methylbutane was used as a demonstration case. The plant was operated continuously in a stable manner for over 50 hours. In this first step the distillate was removed from the system. In the next step the distillate will be redirected into the reactor feed to study the accumulation of impurities.

The pilot plant required very little supervision; mainly the refilling of the feed pump and the weighing of the product containers. The plant was running overnight and over the weekends without problems or supervision. The plant required relatively small space and it was fitted inside a fume cupboard. Thanks to the small size and low supervision requirement, the small scale pilot plant offers the possibility to test processes earlier in a process development project. A number of them could even be operated in parallel by a single operator.

The estimated cost for the whole system was 45 000 €. This estimate includes all the custom made heaters and heat exchangers, the distillation column, instrumentation and measurements, mass flow controllers, pumps, valves, and piping. The cost of labor, and the price of gas chromatographies were excluded.

Acknowledgements

We are grateful to the Finnish Funding Agency for Technology and Innovation for financial support. A.T. Sundberg acknowledges the support of the Fortum foundation. P. Uusi-Kyyny acknoledges the support of the Academy of Finland.

References

1. Gertenbach, D., Cooper, B., **2009**, *Scaleup issues from bench to pilot*, AIChE annual meeting, November, Nashville, USA

2. Lowenstein, J., **1985**, *The pilot plant*, Chem. Eng. 9, 23, 62-75.

3. Mannan, S.(ed), **2005**, *Lee's Loss Prevention in the Process Industries*, 3rd ed., Vol. 1, Elsevier 8/13, A10/2-A10/8.

4. Sundberg, A., Uusi-Kyyny, P., Alopeus, V., **2009**, *Novel micro-distillation column for process development*, Chem. Eng. Res. Des. 87, 705-710.

5. Wörz, O., **1995**, *Process development via a miniplant*, Chem. Eng. Process. 34, 261-268.

Experimental Investigation of the Scale up of Turbula Mixers on the Basis of Kinematic and Dynamic Similarities

Claire Mayer-Laigle, Cendrine Gatumel and Henri Berthiaux,
Université de Toulouse ; Mines Albi ; CNRS ; Centre RAPSODEE, Albi cedex 09, France

Introduction

Many processes from a wide range of industries (cosmetic, pharmaceutical, food) use powders as raw materials. In most of these, mixing or premixing of dry particles is a critical step, the mixture quality being known to have an influence on the taste of a food product or the bioavailability of a drug. To meet the specifications of the finished product, the choice of the mixing technology and its operating conditions is therefore a key factor for the process.

However, the complexity of granular materials leads to difficulties in predicting the behavior of powders inside a blender. Most of the time, the analysis and the optimization of a mixing process for a given mixture cannot be performed without an experimental work that is usually tedious. The cost and the quantity of raw materials needed to conduct these studies imply that such work is usually done at the laboratory scale. The main challenge lies then in the extrapolation of results from lab scale to pilot or industrial scale with a small number of tests.

In basic chemical engineering, scale-up is envisaged on process similarities between different scales and employ dimensional analysis. Two processes may be considered completely similar if they take place in a similar geometrical space and if all the dimensionless numbers used to characterize the process are equal [1]. Usually, it is impossible to reach the complete similarity, especially with granular materials for which a classical dimensional analysis is arduous to conduct [2]. Indeed, on one hand it is hard to determine the impact of each factor influencing the flow of a given powder mixture and on the other hand, these factors are difficult to quantify.

Thus, for powder mixture, the principle of similarity is applied without accurate knowledge of dimensionless number describing the system. In this case, similarities could be geometric, kinematics or dynamic.

Geometric similarities refer to a constant ratio between all linear lengths of the prototype and the extrapolate system. In practice, suppliers try to abide by the geometric similarity and they are generally observed.

Kinematic similarities: Two systems will be considered similar from the kinematic viewpoint if two corresponding points will undergo a similar trajectory within a similar time. If times and trajectory covered are respectively on a ratio of τ and λ, the corresponding velocity and acceleration will be on a ratio of $\lambda.\ \tau^{-1}$ and $\lambda.\ \tau^{-2}$. For a liquid, this means that fluid streamlines are similar. For a powder, we are not talking about streamlines but about flow regimes linked to the velocities of particles [3]. Thus, it is assumed that kinematic similarities are complied if particles have the same velocity at each mixer scales ([4], [5]).

Dynamic similarities are based on keeping the same ratio between forces acting on corresponding points in each system. Dynamic similarities as often linked to Froude number resulting in a ratio of the inertial forces to the G-forces. In a blender, inertial forces boil down to centrifugal forces. If the use of the Froude number is widespread to extrapolate a blender, its use as a single dimensionless number has not been proven ([2], [6]).

This paper deals with the scaling-up of Turbula mixers, commonly used in industry and research. This works is based both on geometric, kinematic and dynamic studies performed thanks to a Solidworks simulation for three scales of Turbula, and experimental results obtained in the differents scales for two different powder systems.

Materials and Methods

The Turbula mixer is used for the homogeneous mixing of powder substances. The efficiency of this mixer is based on the interaction of revolution, translation and inversion as per the geometric theory developped by Schatz. The product is mixed in its own closed container which is set into three dimensional movement that exposes the product to always changing, rhythmically pulsing motion. This resulting nearly chaotic motion allows to obtain a homogeneous mixture rapidly for many granular systems.

Description of the mixing device and geometric study

This device consists of a container in a mixing basket set by 2 stirrups at 2 rotary axes. One of them drives the whole system by rotating at a fixed speed which can be adjusted (4 or 5 possible speeds according to the different scales). The movement of the system is possible if and only if the radius of the stirrup is equal to the center distance between the two stirrups [7]. The mixing basket can hold any form of container. In this study, we will only use the maximum size of standard container provided for the equipment.

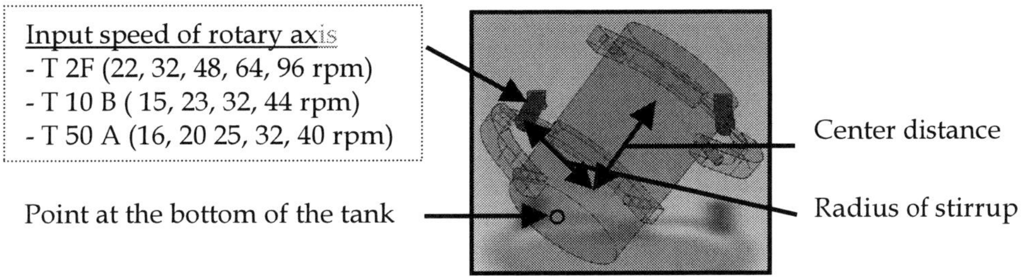

Figure 1 : Schematic representation of a Turbula mixer

Three Turbula scales were investigated, namely laboratory Scale – Turbula T 2F, pilot scale – Turbula T 10 B, industrial scale – Turbula T 50 A. All characteristic dimensions of this three Turbula scales are summarized in table 1

Table 1: Characteristic dimensions for the Turbula T 2F, T 10B et T 50A

	Laboratory Scale	Pilot scale : T 10 B		Industrial scale : T 50 A	
			T 10 B/ T 2F		T 50 A/ T 2F
Volume of the container	2 L	17 L	8.5	55 L	27.5
Radius of the stirrup	130 mm	225 mm	1.73	360 mm	2.77
Diameter of the vessel	130 mm	252 mm	1.93	370 mm	2.84
Length of the vessel	200 mm	360 mm	1.8	565 mm	2.83

Gobally, the characteristic lengths of a Turbula mixer appear to comply with geometric similarities. We could still remark a slight distortion on diameter of the vessel for the pilot scale mixer. When a recipient of another dimension is inserted in the basket, this distortion could be more important

Kinematic and Dynamic study of Turbula mixer

Depending on the input speed of the rotary axis and on the size of the mixer, acceleration and velocity for points of the vessel are very different. These kinematic and dynamic criteria will probably influence the quality of the resulting mixture.

Each of the three scales of turbula has been modeled with Solidworks. Maximal amplitude velocities and accelerations for a reference point at the bottom of the tank was extracted from simulations and correlation tables of input speed were built on a kinematic criterium (equality of velocity for the reference point) or a dynamic criterion (equality of Froude number) (see figure 2). We assume that this point sum-up velocities and accelerations in the powder bulk. Indeed, This corresponds to the bottom of the powder bed inside the container. During the mixing, the powder is "reversed". Thus, particles that were at the bottom of the bed will be at the top and vice-versa.

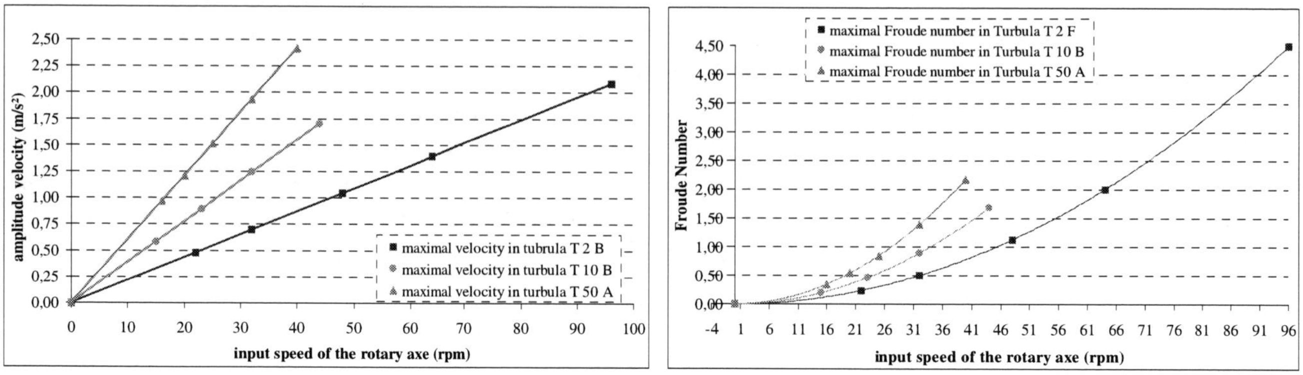

Figure 2 : Corresponding input speeds in each turbula size according to a kinematic (left) or dynamic (right) criterion

From Figure 3, velocities at the bottom of tanks obtained when using the T 2F at 32 rpm are equivalent to those obtained in a T 10B with an input speed of about 17 rpm. The same identical acceleration will be obtained within the T 2F and T 10 B for speeds respectively 32 and 24 rpm.

Note that the Froude number in the different mixers ranges from 0.5 to 2.5, excepted in Turbula T 2F where it reaches 4.5 for the imput speed of 96 $tr.min^{-1}$. For a Froude number <<1, G-forces predomine, and for a Froude number >> 1, centrifugal forces are preponderant.

Characteristics of powder mixtures studied

In this study, two binary mixtures were used. The first one is is a *theoretical mixture (TM)* composed of lactose (85 %) and semolina (15 %). The second is an *industrial mixture* used to manufacture bipolar plates for fuel cells industry. It consists of 85% graphite and 15% of thermosetting polymer. This system will be called *industrial mixture (IM).* Particle size distribution characteristics, as well as Carr and Hausner indexes for the two mixtures are summarized in table 2.

Table 2 : characteristics of powders and flow properties

	TM : theoretical mixture		IM : industrial mixture	
	component 1	component 2	component 1	component 2
Volume mean diameter	65.4 μm	≈ 1 mm	75.7 μm	175.3 μm
Span (D 10% -D 90 %)	1.952	0.723	2.657	3.400
Carr Index	38,10		44,4 %	25,8 %
Hausner Index	1,61		1,79	1,35

The quantitative characterization of mixture quality requires the prior set-up of a sampling procedure (number and size of samples, “how to withdraw them ?”), as well as the use of a reliable method of measurement. There is no universal method of mesurement: it must be adapted to the nature of the powders as well as the intended use of the final product.

Method of measurement for the theoretical mixture

The *theoretical mixture* is discharge through a funnel on a conveyor belt. An inclined piece of stiff cardboard allows the formation of a powder layer of constant width and thickness (see figure 3.a).

Lactose, is a white powder and semolina was coloured in black by iodine adsorption. This color difference could be detected easily by an image processing technique. The measurement chain consists of a linear CCD camera placed over the conveyor belt which transports the loose materials, as well as a PC computer. A time sequence of images is captured by the camera and represents the solids mixture on the belt during conveying. These images are treated with a program developed on the labview software to determine the composition of each sample. The pixel size roughly corresponded to 0.6 g of mixture.

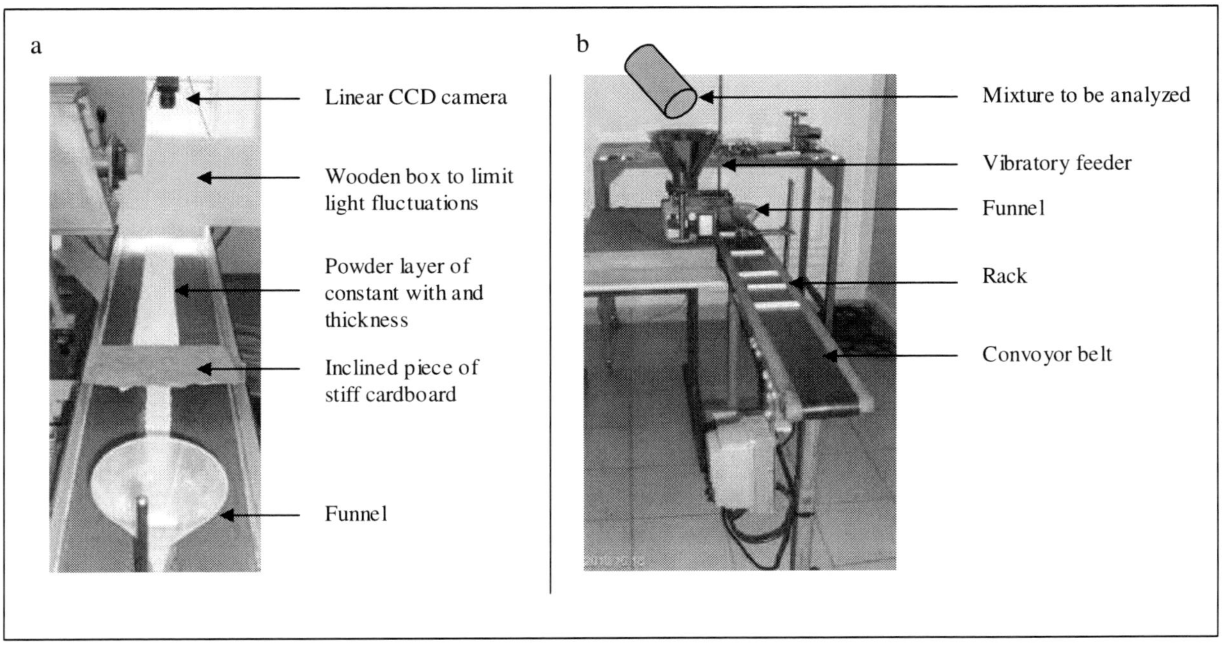

Figure 3 : Device for analyzing composition of the theoretical mixture (a) and sampling procedure for the industrial mixture (b).

Method of measurement for the Industrial mixture

As for the theoretical mixture, powder is discharged via a vibratory feeder on a conveyor belt whose speed is adjustable. No online method of measurement was available for this mixture, so an exhaustive sampling procedure was not an option. In regard of the final use of the industrial mixture, it was determined that the scale of scrutiny must be about 70 mg. Filling rate of the 3 vessels was set to 50 % by volume corresponding to mixing masses of 420 g, 3.570 kg and 11.550 kg or 6000, 51 000 and 165 000 potential samples respectively. A statistical analysis, based on the assumption that the compositions of samples are distributed according to a normal distribution showed that 60 samples were required to compare two different mixtures. Samples are withdrawn randomly thanks to racks (see figure 3.b). By adjusting the speed of the conveyor and the amplitude of vibration, mass of sample have been set to 70 mg.

The method of mesurement used is based on the conbustion of the thermosetting polymer in a oven at 550°C during 2 hours. TGA confirmed that all the polymer was burned without residual ash. A loss weight of 0.4 % was also oberved for the graphite because of some moisture adsorbed on the surface of graphite particles. Composition of graphite inside the sample is determined as the ratio between the corrected mass of graphite (to account for the weight loss of 0.4%) and the sample mass.

Mixture homogeneity was assessed by the use of the intensity of segregation [8] derived from the coefficient of variation of the compositions of the samples. A decrease in the intensity of segregation reflects an increase in the homogeneity of the mixture:

CV = σ/μ (with σ: standard deviation and μ: average composition of the samples analyzed)

Results and discussion

Mixing kinetics in turbula T 2F

Figure 4 shows the kinetics of mixing in a Turbula T 2F for *the theoretical mixture*. The evolution of the intensity of segregation is represented as a function of the number of revolutions. The points shown on the curve correspond to an average of 3 to 5 measurements carried out as part of a reproducibility study. In the early stages of mixing, the relative varition on the CV could rise up to 30 % . As an example, for 3 revolutions at 32 rpm, the CV varies between 3.63 % and 4.77%. For longer mixing time, it lays close to 7 %. For this type of mixture, there may be a significant impact on mixture quality of mixer filling, leading a certain compaction of the powder before mixing. Thus, the interpretation of the first points of the curve is rather difficult. In addition, for short mixing times and low input speed (22 rpm and 32 rpm) the values of CV obtained either stagnate (22 rpm) or oscillate (32 rpm) before starting to really decrease. Table 3 shows the values of CV obtained for the different tests performed.

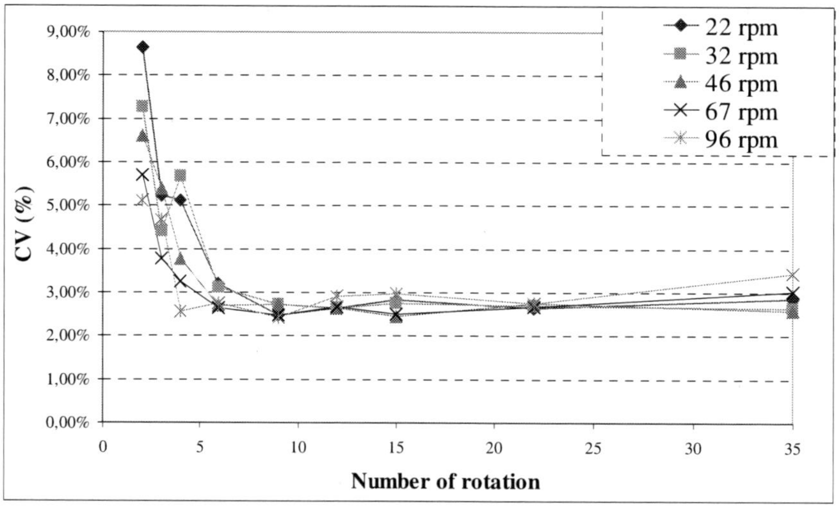

Figure 4 : Mixing Kinetics in Turbula T 2F for the theoretical mixture

Table 3 : CV obtained for different test carried out at small numbers of revolutions

	CV obtained for 3 revolutions at 22 rpm	CV obtained for 4 revolutions at 22 rpm	CV obtained for 4 revolutions at 32 rpm	CV obtained for 4 revolutions at 32 rpm
Test 1	5.16 %	4.71 %	4.40 %	5.40 %
Test 2	5.29 %	5.70 %	4.48 %	5.87 %
Test 3		4.99 %	4.77 %	5.70 %
Test 4			3.63 %	
Test 5			4.74 %	

For higher input speeds, the CV values decrease more evenly. All kinetics seems to reach the same value for 22 revolutions before spreading as a function of speed. The higest speeds lead to higher values of CV. The tests carried out for long mixing time (about 500 revolutions) confirm this tendency. (see table 4)

Table 4 : CV obtained for different test carried out at 35 and 500 revolutions

Input speed	CV for 35 revolutions	CV for about 500 revolutions
22 rpm	2.87 %	2.33%
32 rpm	2.63%	2.16%
46 rpm	2.60%	2.43%
67 rpm	3.03%	2.55%
96 rpm	3.44%	3.06%

It is well known that the mixing patterns obtained results from the effects of mixing and segregation mechanisms. According to the differents forces experienced by the powder in the mixer and its flow properties, some mechanisms will be predominant over others.

At low input speeds, the Froude number is less than 0.5 and G-forces are principally responsible for the mixing state. At higher speeds, the Froude number is greater than 1 (up to 4.5 for 96 rpm) and centrifugal forces prevail. The model mixture has a strong disparity in terms of particle size and particle density and is therefore especially prone to segregate (see table 1).

In the early stages of mixing, at low input speed, G-forces may lead to particle segregation. These effects are more or less offset by mixing mechanisms. This may explain the fluctuations observed. When speed is more important, the role of G-forces is less, and the effects of the associated segregation are not visible. By contrast, the dispersion observed for longer mixing times may result from a reorganization of particles in the mixture due to centrifugal forces.

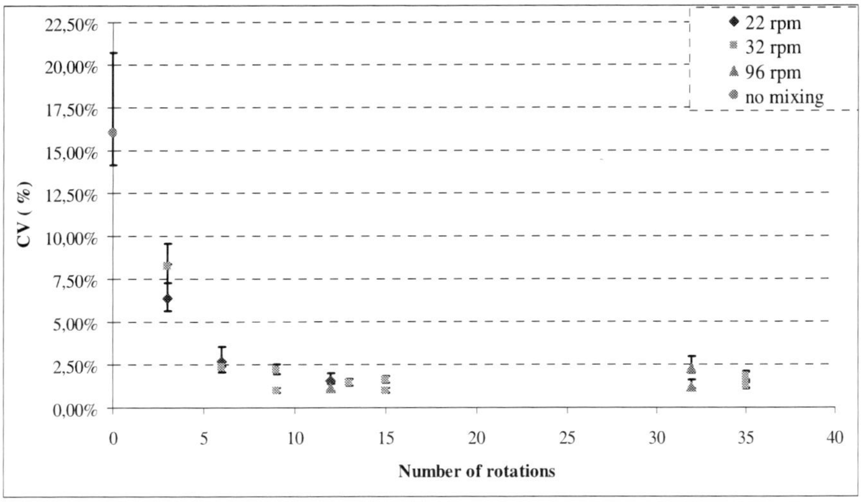

Figure 5 : Mixing kinetics in Turbula T 2F for the industrial mixture

Kinetics of mixing obtained in Turbula T 2F for the industrial mixture are shown in figure 5. The errors bars have been determined by Student's t-test and represent errors introduced by sampling. As the number of samples have been fixed, these are all the more important when mixtures are of a low quality. For a number of revolution greater than 12, there is no distinction between the CV, whatever the input speed considered. Thus, only the first points of the kinetic curve seems to depend on the input speed and the number of revolution.

Scale-up

We focused, in this study of scaling-up, on the quality of the mixture obtained at the output of the mixer. We sought to define which critirion (kinematic or dynamic) allows to obtain a mixture of the same quality (i.e equality of CV) for the same number of revolutions. From the graph presented in figure 2, equivalent input speeds according to kinematic or dynamic criteria were determined. In the light of the previous mixing kinetics, we decided to compare values of CV obtained from tests performed for 500 revolutions in turbula T 2F and Turbula T 10 B for the model mixture (figure 6) and tests performed for a small number of revolution (less than 12) for the industrial mixture in the three Turbula sizes (see figure 7 and 8).

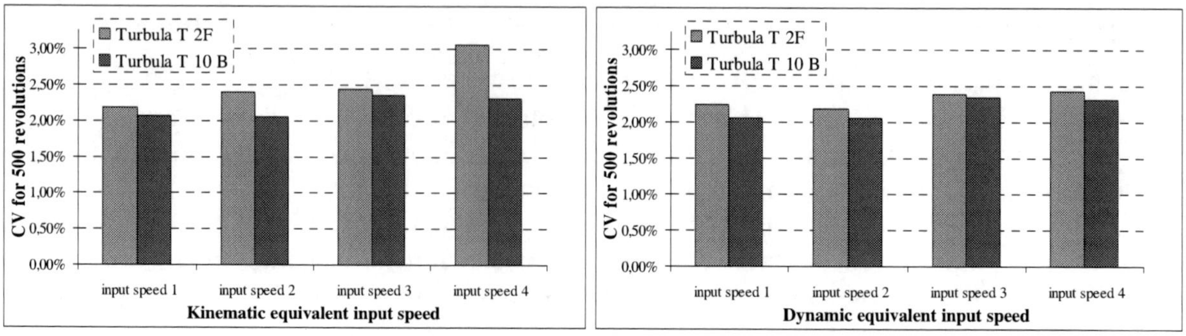

Figure 6 : Comparison of the CV obtained for kinematic equivalent input speed(left) or dynamic equivalent input speed (right)for the theoretical mixture

For the *theoretical mixture,* it seems that the dynamic criterion allows to obtain CV values closer than for the kinematic criterion, especially for the equivalent input speed 2 and 4. Furthermore, whatever the tests performed, the mixtures obtained in the Turbula T 10B seems to lead to a higher quality of the mixtures. These results are in concordance with those obtained by Cleary and Sinnott [6] in a V-blender. They demonstrated that the quality of the mixture obtained in the pilot-scale mixer for the same Froude number and number of revolutions is better than that obtained in the laboratory scale mixer.

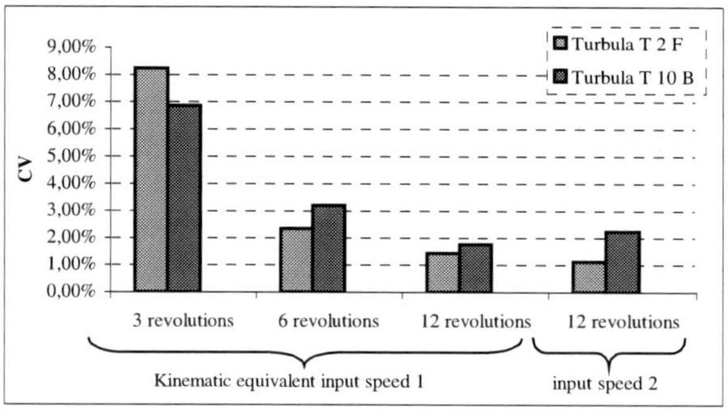

Figure 7 : Comparison of the CV obtained for kinematic equivalent input speeds for the IM

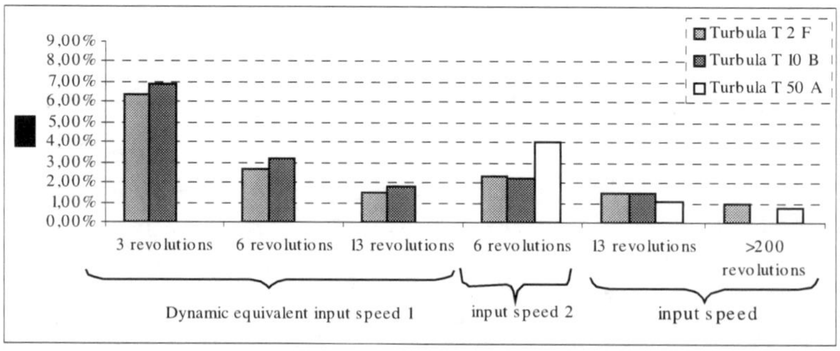

Figure 8 : Comparison of the CV obtained for dynamic equivalent input speeds for the IM

In the *industrial mixture(IM)* too, the difference betwen the CV values obtained in the three sizes of Turbula are lower if the scale-up is based on a dynamic criterion. In figure 8, it can be noted that for 6 revolutions at the input speed 2, the CV value obtained in the Turbula T 50A is higher. This can be explained by the inertia of the mixer: to reach the specified speed, the mixer needs a certain number of revolutions. This will be all the more important if the speed of the mixer is high. It also explains why tests for 6 revolutions for the input speed 3 have not been performed. Even if it did not seem obvious on the mixing kinetics, it appears that greater input speed leads to better mixing qualities for a long mixing time (number of revolution > 200).

Conclusion

Geometric, kinematic and dynamic studies were performed thanks to a Solidwoks simulation carried out for the three scales of Turbula. From these, kinematic and dynamic criteria for scaling-up Turbula mixers were suggested on the basis of equal velocities or Froude number for a reference point at the bottom of the tank. The mixing kinetics for two powder systems have been built in Turbula T 2F to determine which experimental trials have to be performed in the three Turbula scales to test the criteria. In each case, the intensity of segregation was compared for a same number of revolutions. The dynamic criterion appears to lead to a better correlation between results obtained in terms of homogeneity of powder mixtures, whatever the powder system considered. These results are consistent with those observed in a high shear mixer [4]. Moreover it seems that for the same Froude number and number of revolution, especially for the *theoretical mixture*, the quality of the mixture is slightly higher in the largest mixer. This can be explained by the presence of greater distances to be traveled by the particles leading to a greater homogeneity for the same mixing time. To complete this study, our future work will focus on the determination of mixing mechanisms occuring in the different mixer scales as a function of the input speed. This can be performed thanks to the use of the scale of segregation [9].

The authors acknowledge the french agency for research, ANR, for his financial support through the MASCOTE project.

References

[1] H. L. Langhaar. *Dimensional Analysis and Theory of Models*. Huntington (N. Y.), 1980
[2] Y. L. Ding, R. N. Forster, J. P. K. Seville and D. J. Parker, 2001, Scaling relationships for rotating drums. *Chemical engineering science*, 56:3737-3750
[3] B. C. Hancock et al., "Development of a robust procedure for assessing powder flow using a commercial avalanche testing instrument," *Journal of Pharmaceutical and Biomedical Analysis*, vol. 35, no. 5, pp. 979-990, Sep. 2004.
[4] A. Alexander, T. Shinbrot, and F. J. Muzzio, 2002, Scaling surface velocities in rotating cylinders as a function of vessel radius, rotation rate, and particle size. *Powder Technology*, 126(2) :174–190
[5] H. Nakamura, Y. Miyazaki, Y. Sato, T. Iwasaki, and S. Watano, "Numerical analysis of similarities of particle behavior in high shear mixer granulators with different vessel sizes," *Advanced Powder Technology*, vol. 20, no. 5, pp. 493-501, Sep. 2009.
[6] P. W. Cleary and M. D. Sinnott, "Assessing mixing characteristics of particle-mixing and granulation devices," *Particuology*, vol. 6, no. 6, pp. 419-444, Décembre 2008.
[7] K. Wohlhart, 1981, A Dynamic analysis of the Turbula, *International Symposium on Gearing & Power Transmissions*, Tokyo.
[8] P. Danckwerts, 1952, The definition, measurement and some characteristics of mixtures, *Applied scientific research*. Section A, 3, 279-296
[9] J. Gyenis, "Assessment of mixing mechanism on the basis of concentration pattern," *Chemical Engineering and Processing*, vol. 38, no. 4-6, pp. 665-674, Sep. 1999.

(83a) Intensify Process by Efficient Micromixing

Bharat Marwaha[1], Maria Krusche[2], Frank Schael[2], Kathrin Rodermund[3]and Marcus Grnewald[4], (1)Bayer Technology Services Americas, a division of Bayer Business and Technology Services, LLC, Baytown, TX, (2)Ehrfeld Mikrotechnik BTS GmbH, 55234 Wendelsheim, Germany, (3)Ruhr-Universitt Bochum, Bochum, Germany, (4)Chair of Fluid Process Engineering, Ruhr-Universitt Bochum, Bochum, Germany

Intensify Process by efficient micromixing

Kathrin D. Rodermund[1], Marcus Grnewald[1], Frank Schael[2],
Maria Krusche[2],Bharat Marwaha[3*]

1. Ruhr-Universitt Bochum, Chair of Fluid Process Engineering, Universittsstrae 150, D-44801 Bochum, Germany

2. Ehrfeld Mikrotechnik BTS GmbH, Mikroforum Ring 1, D-55234 Wendelsheim, Germany

3. Bayer Technology Services Americas, Baytown, TX

****Speaker***

Micro process technology enables new pathways for the development of economical, innovative and intensified processes in the chemical industry. Modular Microreaction technology provides clear advantages over conventional process technology through a.)short mixing time, b.) high heat transfer rates, c.) inherent safety of small hold-up, d.) precise control of reaction time and e.) faster scale-up/time-to-market.

The intensified mixing offered by microchannel devices can aid process intensification for those applications where fast reactions are limited by slow mixing time. The resulting process intensification leads to a safer and sustainable process with higher yield, higher selectivity. Quantification of mixing quality can guide selection and feasibility of optimal mixing structure for a particular application.

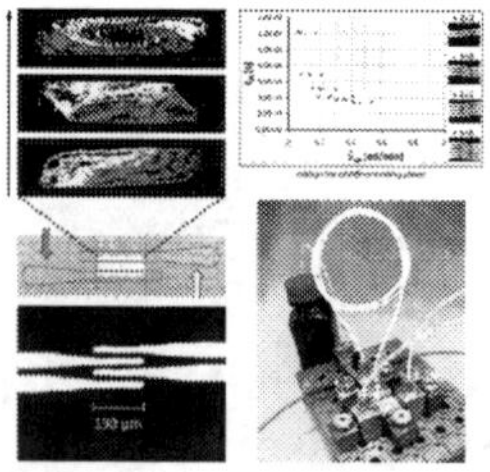

Results of the development of experimental methods for determination of mixing times of micro mixers will be presented. The aim of our research is to develop objective assessment criteria that are independent of the used characterization method. In the current work, the Villermaux-Dushman method is employed to quantify the efficiency of a micromixer . Results show that the peripherals have a significant effect on the mixing process and thus its effects have to be considered in the case of scale-up.

The presentation will also showcase a variety of micromixers, show concept of scale-up and share examples of their applications.

(83b) Economical Comparison of Compact Heat Exchange Reactors for a Small Fischer Tropsch Synthesis Plant

Hani Gadalla, Zhijun Jia and Steven J. Vallee, Chart Energy & Chemicals, La Crosse, WI

The need is growing for small scale Fischer Tropsch plants (1-500 bpd) to produce hydrocarbon products as fuel, fuel additives or synthetic oils and waxes. Fischer Tropsch synthesis may be used for a wide choice of technologies and feedstocks: gas-to-liquid, coal-to-liquid, biomass to liquid and even waste-to-energy. Conventional reactors, such as slurry bed and shell-and-tube, are utilized in large scale gas-to-liquid plants. However, these reactors and many of the auxiliary equipment associated with them are not economically feasible to scale down to the 500 bpd plants. Compact heat exchange reactors (CHER) have been proven as viable reactors for the small scale FT synthesis. Based on the compact heat exchange technology, where small channel sizes and high exchanger metal density greatly increase the heat transfer, CHER can control the reaction tightly and increase the production rate per reactor volume. CHERs are easily scalable by adding more process and coolant layers. Furthermore, by incorporating the CHER in a modular design, plant capacity can be easily increased with the addition of more modules, allowing the plant to expand to 2,000 – 5,000 bpd. An economic analysis for the small scale FT plant will be performed to compare the capital and operating costs of plants with different reactor types, as well as a study on the return of investment for each of the plants.

(83c) Aryl Addition Chemistry on Scale: A Case for the Application of Flow Chemistry to Minimize Unexpected Impurities

Michele Drexler[1], Nga Do[2], Matt Weekly[2], Mark Olivier[2]and Jade Nelson[2], (1)CRD - Engineering Technologies, Pfizer Global Research and Development, Groton, CT, (2)Pfizer Global Research & Development

Aryl Addition Chemistry on Scale: A Case for the Application of Flow Chemistry to Minimize Unexpected Impurities

Nga Do, Michele T. Drexler*, Mark Olivier, Matt Weekly, and Jade Nelson

Chemical Research & Development

Pfizer Global Research & Development

Groton, CT

During evaluation of the key aryl addition for scale of a pharmaceutical intermediate, two impurities were observed during the initial halogen-metal exchange reaction of the aryl bromide**1**with***n***-BuLi in 2-Me-THF solvent. These impurities (**3**and **4**) were observed at combined levels up to 50% (UPLC) area and were evident at all temperatures between -20°C and -78°C. To minimize these impurities by providing more efficient heat and mass transfer for the system, flow chemistry was explored. This presentation will discuss the optimization of the halogen-metal exchange portion of the aryl addition chemistry and the application of a continuous reaction setup to provide a process for scaling the pharmaceutical intermediate.

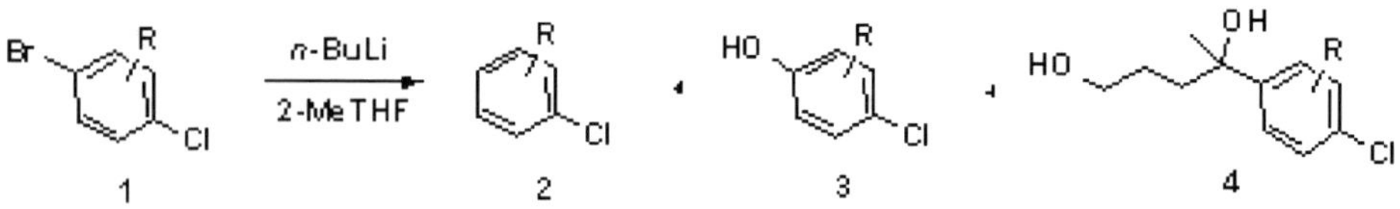

(83d) Application of Radiofrequency Heating for Process Intensification in Pd Catalysed Hydrogenation Over Composite Magnetic Microparticles

Thomas K. Houlding, Pengzhao Gao and Evgeny V. Rebrov, School of Chemistry & Chemical Engineering, Queen's University Belfast, Belfast, United Kingdom

Application of radiofrequency heating for process intensification in Pd catalysed hydrogenation over composite magnetic microparticles

Thomas K. Houlding, Pengzhao Gao, Evgeny V. Rebrov

School of Chemistry & Chemical Engineering, Queen's University Belfast, Stranmillis Road,BT9 5AG,Belfast,UK

Recently, much research in process intensification has been focussed on the development multi-functional reactors, improving processes by system integration [1]. The use of NiZn ferrite nanoparticles embedded in a mesoporous microparticles for radiofrequency (RF) heating, magnetic separation and enhancement of mass transfer in the laminar flow within a continuous flow system provides a novel process intensified platform for system integration. The objective of this study was to explore novel Pd-based supported catalysts for selective and economically viable fine chemical synthesis.

The composite titania microparticles were prepared by sol-gel method from titania precursor in the presence of a structure directing agent and NiZn ferrite nanoparticles (grain size 15-25 nm) in the sol following earlier developed methodology [2,3]. After calcination at 300°C, the porous network was filled with a Pd precursor by incipient wetness impregnation. Then the microparticles were dried, calcined and reduced in hydrogen flow.

The heating rate of composite microparticles in the AC field (100-400 kHz, current 20-100 A) was determined of powders and slurries containing the desired amount of solid material. Induction coils with either 4, 5 or 6 turns were used as applicators to generate RF fields with different frequencies. During operation the induction coil was cooled from inside by circulating water at 10°C. The heating rates were studied as a function of ferrite composition (Ni and Zn content), the total loading of ferrite onto titinia microparticles as well as of the field strength and frequency. The temperature was measured with a fibre optic probe. The highest heating rate was observed over microparticles containing NiZn ferrite nanoparticles with low coercivity, with average powder heating rates of up to 40 K/min. Specific absorption rates (SAR) of up to 10 W/g were observed for composite microparticles containing magnetite nanoparticles.

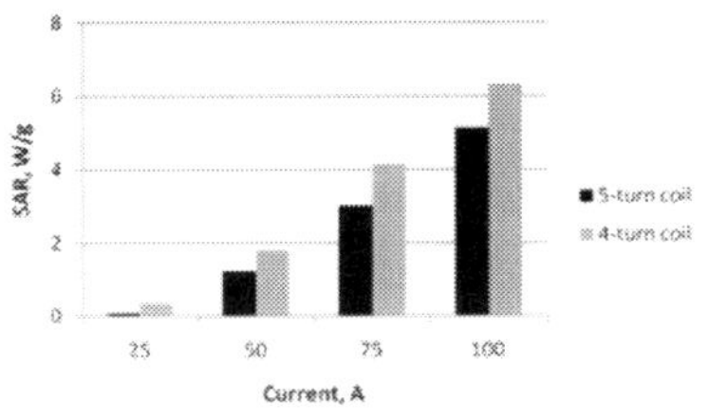

Figure 1. Specific absorption rate of SiO_2@magnetite microparticles as a function of current and coil type

The catalytic activity of composite catalysts in hydrogenation reactions was determined in a reactor made of PEEK at a total H_2pressure of 3 bar. Reference experiments were carried out using conventional heating with a thermostat. During experiments under RF-heating, the liquid temperature remains considerably below the catalyst temperature, which was estimated from the Arrhenius plot for the reaction rate of several selected transformations.

References

[1] Hessel V. Novel Process Windows - Gate to Maximizing Process Intensification via Flow Chemistry. Chem. Eng. Technol. 2009;32:1655-1681.

[2] Rebrov EV, Berenguer-Murcia A, Skelton HE, Johnson BFG, Wheatley AEH, Schouten JC. Capillary microreactors wall-coated with mesoporous titania thin film catalyst supports. Lab Chip. 2009;9:503-506.

[3] Protasova LN, Rebrov EV, Skelton HE, Wheatley AEH, Schouten JC. A kinetic study of the liquid-phase hydrogenation of citral on Au/TiO2 and Pt-Sn/TiO2 thin films in capillary microreactors. Appl. Catal. A. 2011;399:12-21.

Rotating Zigzag Bed Application in Extractive Distillation process of THF-Methanol-Water system

Zhi-chao Xu[1], Jian-bing Ji[2], Guang-quan Wang, Xiao-hua Li, Yu-min Li
Zhejiang University of Technology, Hangzhou, Zhejiang, P. R. China

Abstract

Rotating zigzag bed (abbreviated to RZB) is a kind of high-efficiency gas-liquid mass transfer device and about 300 units have been applied in industrial distillation as well as stripping processes. This paper presents its application in extractive distillation, which involves the separation of THF-methanol-water system. Process parameters were obtained using Aspen simulation software and according to these parameters the RZB was designed. Industrial practice shows that methanol residue in THF product is only 24ppm. The height and the volume of RZB is only about 1/7 and1/2 compared with the packed column

Key words: Rotating zigzag bed, Higee, Extractive distillation

Rotating bed, as a kind of process intensification equipment, was first developed in 1980s[1] and from then on, various rotating beds with different rotor structures were proposed by researchers.[2] These rotating beds have been commercialized in VOC removal from groundwater,[3] deoxygenization of oilfield injected water,[4] nanoparticles preparation,[5] waste gas purification[6] and reactive separation.[7] However, studies on rotating bed in distillation processes were relatively few and industrial applications were even less.[8] Since first successfully commercialized in 2004, rotating zigzag bed (abbreviated to RZB) has found about 300 application cases in distillation and stripping processes, with excellent economic and social benefits. This paper was focused on a pharmaceutical plant, where THF-methanol-water system was required to separate in limited area and space. Due to less height and smaller volume, RZB was qualified for this task. The boiling point temperature between THF and methanol was only 1.5℃ and they also form a homogeneous azeotrope. Generally it is very difficult to produce high-purity THF, e.g. methanol residue less than 100ppm, using conventional distillation. Therefore extractive distillation was employed and a unit of RZB was designed and applied with a satisfactory result.

1. Design parameters selection by Aspen Plus

For the design of extractive distillation, the first and important thing is selection of a proper extractant. Water and ethylene glycol were good alternatives for this system because they can associate with hydrogen bonds in methanol molecules.[9,10] Due to presence of water in this system itself, water was selected as extractant so as not to introduce additional substance. The mass fraction of feed is methanol 10%, THF 80% and water 10%. The separation requirement is less

[1] Author: Senior engineer. Tel:+86-571-88320936. Fax: +86-571-88320053. E-mail:xuzc@zjut.edu.cn
[2] Corresponding author. Tel:+86-571-88320053. Fax: +86-571-88320053. E-mail:jjb@zjut.edu.cn

than 100ppm methanol in THF product. In Aspen Plus simulation, NRTL was chosen as property package. The simulation flowsheet was shown in Fig.1, where stream 1 was feed, stream 2 extractant, stream 3 distillate product and stream 4 bottom product. The simulation inputs were as follows: the total number of theoretical plates was 18; the feed location of stream1 and 2 was 13th and 5th plate respectively; the reflux ratio was 0.9, and the ratio of extractant to feed was 1. The simulation results were partially given in Table 1. It is noted that the pressures in Table 1 were determined according to the characteristics of RZB.

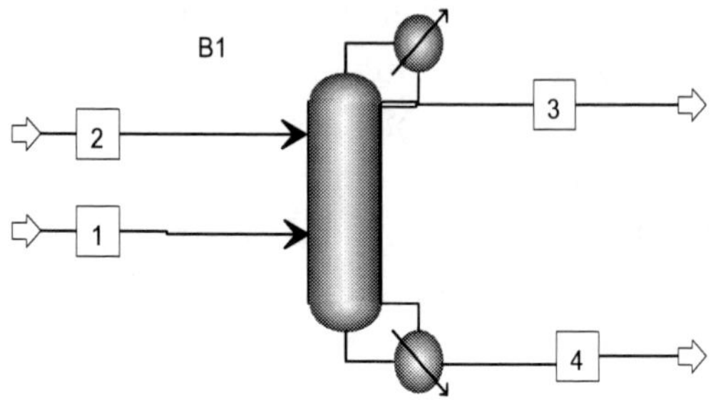

Fig.1 Flowsheet of extractive distillation

Table 1 Results of Aspen Plus simulation

	1	2	3	4
Temperature C	30	63	63.43	92.81
Pressure bar	1.52	2	1.0133	1.0649
Mass Fraction				
METHA-01	0.1	0	3.78E-05	0.086
TETRA-01	0.8	0	0.949	0.003
WATER	0.1	1	0.051	0.910

It can be found from the simulation results that the distillate product has the azeotropic composition of THF and water, with methanol fraction less than 40ppm.

2. Characteristics and feed location distribution of RZB

RZB was a kind of high-efficiency Higee,[2,11,12] the core of which was a rotor combining a rotating disc with a stationary one. Two series of concentric circular sheets were connected with the rotational and stationary disc respectively. These baffles are assembled together with alternate sheets staggered in a vertical direction with respect to adjacent sheets. The feeds were easily configured because the upper part of the rotor was stationary. The main contacting zone in RZB was a pair of baffles, one rotating baffle and stationary baffle adjacent to it. Experiments and industrial practices have shown that a pair of baffles was equivalent to 0.6 theoretical stage.[11]

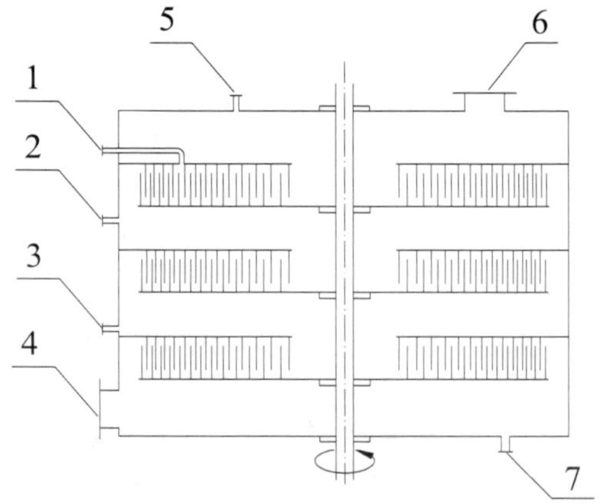

Fig.2 Inlets and outlets of the RZB
1. Extractant inlet 2. Thermometer nozzle
3. Feed inlet 4. gas inlet 5. Reflux inlet
6. Gas outlet 7. Liquid outlet

Fig. 3 The field picture of the RZB

Therefore, it is easy to determine the number of stationary and rotating baffles as well as locations of feed and extractant according to the simulation results.

The desired capacity was 0.5ton feed per hour and, according to gas and liquid loads on each theoretical plate, the designed RZB has three rotors with a diameter of 850mm. The extractant was introduced in the RZB between the stationary baffles in the upper rotor. And the feed entered between the upper and middle rotor. The inlet and outlet arrangement of the RZB was shown in Fig.2. Fig.3 was the field picture of the RZB.

3. Industrial process and operation results

Fig. 4 shows the industrial flowsheet of Higee extractive distillation using RZB, in which the condenser was tailor-made. The vapor flows in the tubes of the condenser and the liquid flows in the shell. The shell was divided into two zones. In the first zone, the liquid was extractive agent—water, which temperature was close to that of the top of the RZB after heat exchange. In the second zone, the cold media was cooling water as usual.

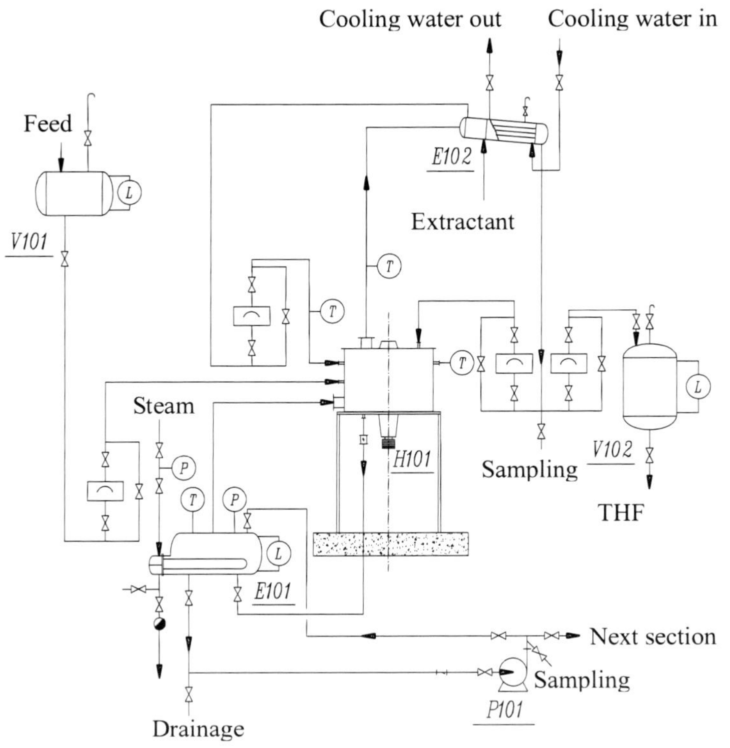

Fig.4 Flowsheet of industrial Higee extractive distillation using RZB.
V101- elevated tank. V102- product tank. H101-RZB. E101- reboiler. E102- condenser. P101-pump.

The feed from an elevated tank entered into the RZB through inlet after measuring the flow rate. The liquid exiting from the RZB returned to the reboiler. The outlet vapor then entered into the condenser and is condensed to liquid. Then the liquid was split into two streams, one was reflux and the other distillate, respectively flowing into the RZB and product tank after measuring the flow rate. The reflux liquid contacted with the vapor from the reboiler and mass transfer was realized. The extractant was heated by condensation of top vapor. The bottom product was pumped to the next section. In order to accurately withdraw the sample from the reboiler, pumparound was performed before sampling.

It might also be noted that at the early stage of operation process, when the temperature measured in the RZB started to increase, the feed valve of extractant should be switched on to avoid the methanol presence in top vapor. Once the methanol entered into the condenser, it would require a long time of reflux to reduce its concentration in the distillate.

When the system was running, measuring data of samples agreed well with designed values. Especially, at total reflux stage, the methanol concentration in the distillate was only about10ppm.

During commission, the actual composition of the feed was THF82.8%, methanol10.5% and water 6.7%. The operation parameters were: feed flow rate 500kg/h, distillate flow rate 432kg/h, reflux flow rate 432kg/h. Analysis results of the distillate were THF94.9%, water 5.1% and methanol 24ppm.

4. Height and volume comparison between RZB and packed tower

If the same extractive distillation task was performed in a packed tower containing Sulzer BX500 structured packing, the tower had a diameter of 0.5m and a height of 10m (exclusive of skirt). The RZB given in Fig. 3 had a diameter of 1m and a height of 1.35m. Table 2 presented the size comparison between the RZB and packed tower. From Table 2, the most obvious difference is that the RZB was far lower than the packed tower. The RZB is of significance for the plants where the area and space was relatively devoid.

Table 2 Size comparison between RZB and packed tower

	Diameter (m)	Height (m)	Volume (m^3)	Height ratio (Packed tower/RZB)	Volume ratio (Packed tower/RZB)
RZB	1	1.35	1.06	7.4	1.85
Packed tower	0.5	10	1.96		

5. Conclusions

Successful applications of the RZB in industrial distillation processes fully demonstrated its advantages such as high gas-liquid mass transfer rate and small size. Because multiple feeds can be easily installed on the RZB, it was applicable to extractive distillation, which requires two middle feeds. In the application case of THF-methanol-water system, the methanol concentration in THF product was only 24ppm, satisfying the requirement (<100ppm) set by the customer.

References

1. C. Ramshaw (1983), "'HIGEE' distillation—an example of process intensification," *Chem. Engr.*, February, pp. 13–14.
2. G Q. Wang, O. G. Xu and Z. C. Xu etc. (2008), "New HIGEE-Rotating Zigzag Bed and Its Mass Transfer Performance," *Industrial & Engineering Chemistry Research,* October, pp. 8840–8846.
3. Surinder P. Singh, James H. Wilson and Robert M. Counce, etc. (1992), "Removal of volatile organic compounds from groundwater using a rotary air stripper," *Ind. Eng. Chem. Res.*, February, pp. 574–580.
4. H. Zhao, L. Shao and J.F. Chen (2010), "High-gravity process intensification technology and application," *Chem. Eng. J.*, February, 588–593.
5. J.F. Chen and L.Shao (2007), "Recent Advances in Nanoparticles Production by High Gravity Technology — from Fundamentals to Commercialization," *J Chem Eng Jpn*, November, pp. 896-904.
6. Liu, Y Z (2009), "Chemical Engineering Process and Technology in High Gravity," *China National Defense Industry Press*, Beijing.
7. Trent D, Tirtowidjojo D and Quarderer G (1999), "Reactive stripping in a rotating packed bed for the production of hypochlorous acid," In: Green A, ed. *3rd International Conference on Process Intensification for the Chemical Industry*, London: BHR Group, pp.217–231.

8. G.Q. Wang, Z.C. Xu and J.B. Ji (2011), "Progress on Higee distillation—Introduction to a new device and its industrial applications," *Chem. Eng.. Res. Des.*, February, pp. 1434-1442.
9. Xu Wen you and Chen Xiao ping (2002), "Recovering Tetrahydrofuran from Waste Pharmaceutical Liquor Using New Separation Methods," *Chemical Engineering (China)*, January, pp.17–19.
10. Zheng Chun zhi, Li Feng and Wen Ying pin, etc. (2010), "Recovery of Tetrahydrofuran and Methanol from Pharmaceutical Wastewater by the Extract ive Distillation," *Chemical World,* December, pp.734–736, 747
11. Xu Zhi chao， Yu Yun liang and Ji Jian bing (2005), "Rotating Zigzag High-Gravity Bed and Its Application to Distillation," *Petrochemical Technology,* Augest, pp. 778–781
12. Wang, G. Q., Xu, Z. C.and Yu, Y. L. etc. (2008), " Performance of a Rotating Zigzag Bed - A New HIGEE," *Chem Eng Process,* November, pp. 2131–2139.

Operations Optimization of Concentrating Solar Power Plants
I: Challenges and Opportunities

I. Salbidegoitia and I. Mendicute
Torresol Energy
Madrid, Spain

G. Deiu, A. Trenchard, J. Visschers and R. Nath
Honeywell Process Solutions
Europe and Americas

Torresol Energy
Honeywell

Agenda

- Introduction
- Operations Optimization
  - Objective
  - Opportunities
  - Challenges
- Conclusions

Introduction: Torresol

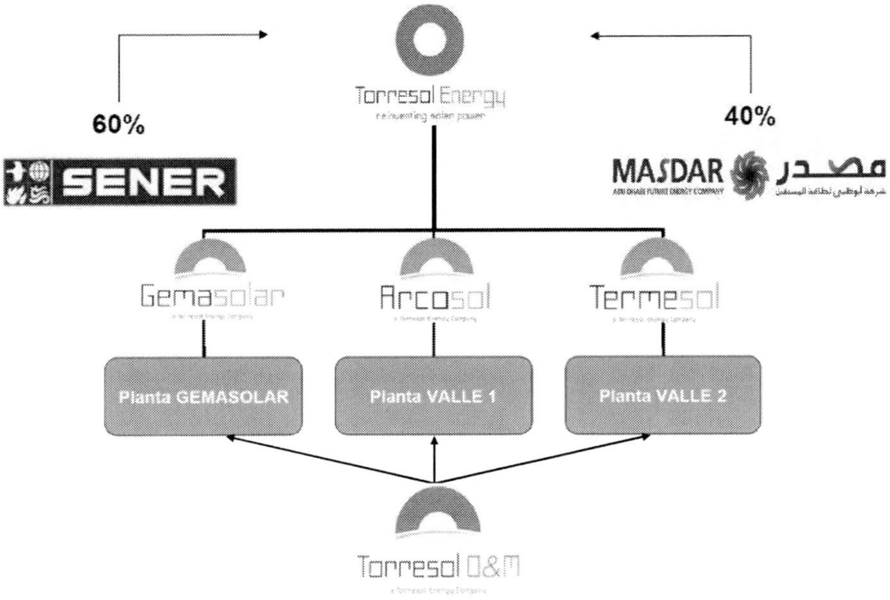

AIChE Spring Meeting
April 3, 2012

Introduction: Solar Energy

- Produced by thermonuclear reaction in the sun
- Produces entire spectrum of EMR

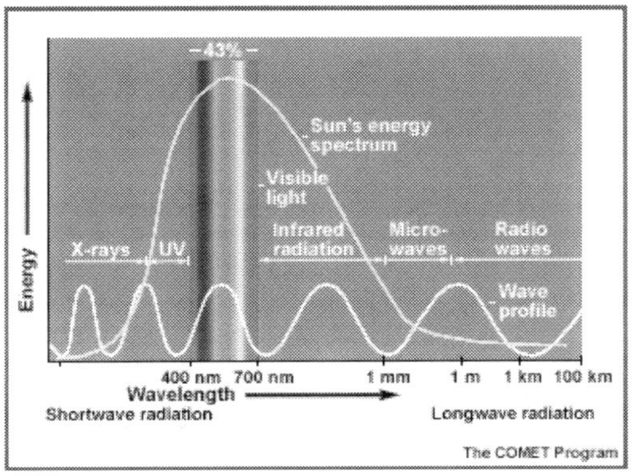

AIChE Spring Meeting
April 3, 2012

Introduction: Solar Energy

- PROs
  - Abundant
  - Clean
- CONs
  - Diffused
  - Cyclical
- Solution
  - Concentration
  - Storage

Introduction: Solar Energy Concentration

- Central or Power Tower
- Parabolic Troughs

Gemasolar

Valle 1 & Valle 2

Introduction: Solar Energy Concentration

AIChE Spring Meeting
April 3, 2012

Introduction: Solar Energy Storage

Valle 1 Salt Storage System

AIChE Spring Meeting
April 3, 2012

Introduction: process diagram

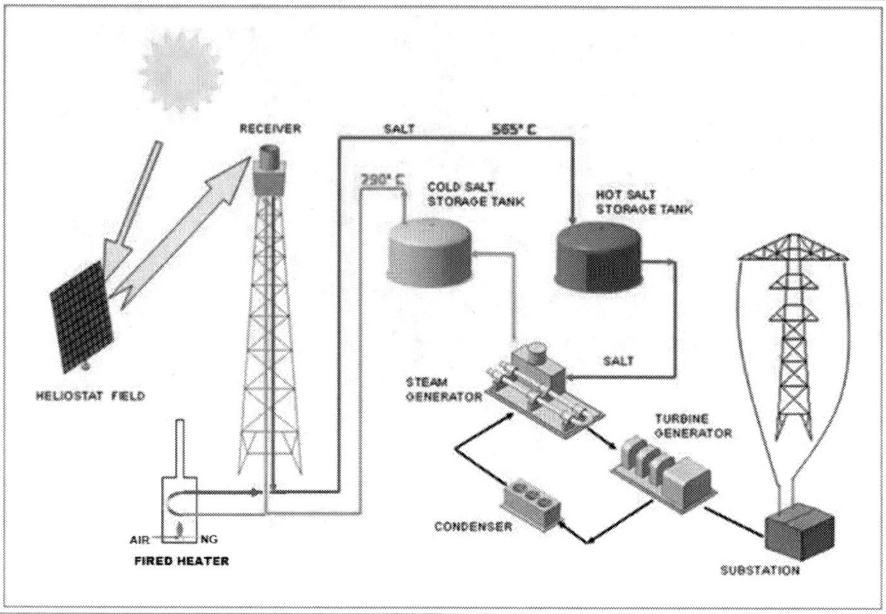

Introduction: economics

High Capital cost, mitigated by

- Federal and state tax incentives
- Green energy pricing incentives
- Use of Fired Heater (limited to an annual maximum)

Operations Optimization – objective

Maximize profit over the optimization time horizon.

$$\approx \text{MAX} \sum_{t=0}^{t=EOY} c\_eee_t * EEE_t + c\_eeeod_t * EEEOD_t$$
$$- c\_eei_t * EEI_t - c\_fei_t * FEI_t - p\_cos * COS_t$$

where

c_eee_t	is the credit for electrical energy export
c_eeeod_t	is the credit for on-demand electrical energy export
c_eei_t	is the cost of electrical energy import
c_fei_t	is the cost of fuel energy import
p_cos	is the cost per transition
EEE_t	is the amount of electrical energy export
$EEEOD_t$	is the amount of on-demand electrical energy export commitment
EEI_t	is the amount of electrical energy import
FEI_t	is the amount of fuel energy import
COS_t	is the indicator for transition, 0/1

AIChE Spring Meeting
April 3, 2012

Operations Optimization – objective

Hour to hour variation

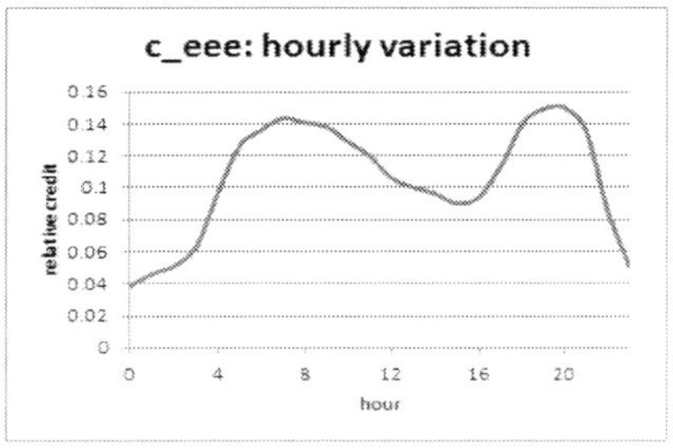

Month to month variation

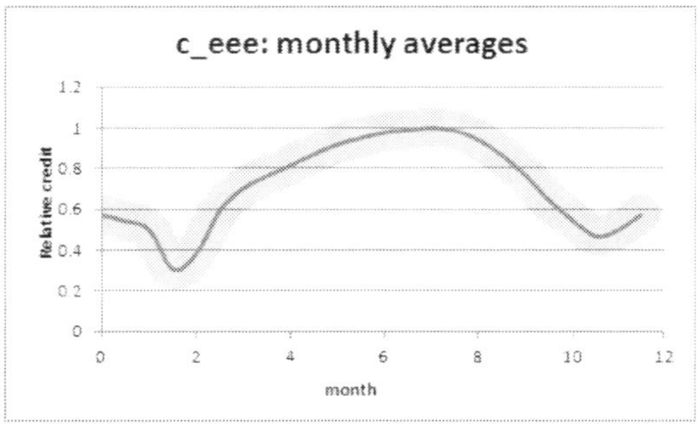

AIChE Spring Meeting
April 3, 2012

Operations Optimization – opportunities

Opportunities:

- Flexibility in production
- Flexibility in use of Fired Heater

Operations Optimization – opportunities (1)

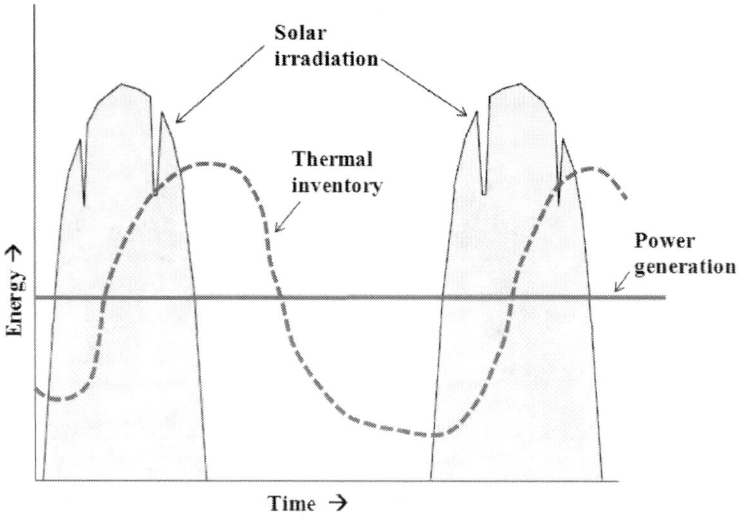

Operations Optimization – opportunities (2)

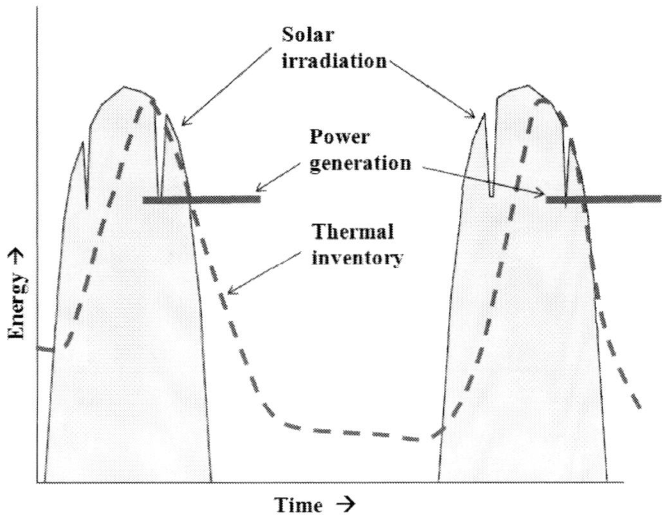

Operations Optimization – opportunities (3)

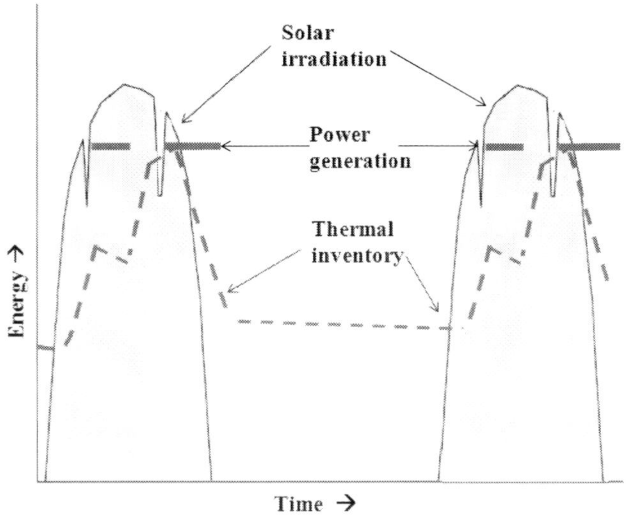

Operations Optimization – opportunities (4)

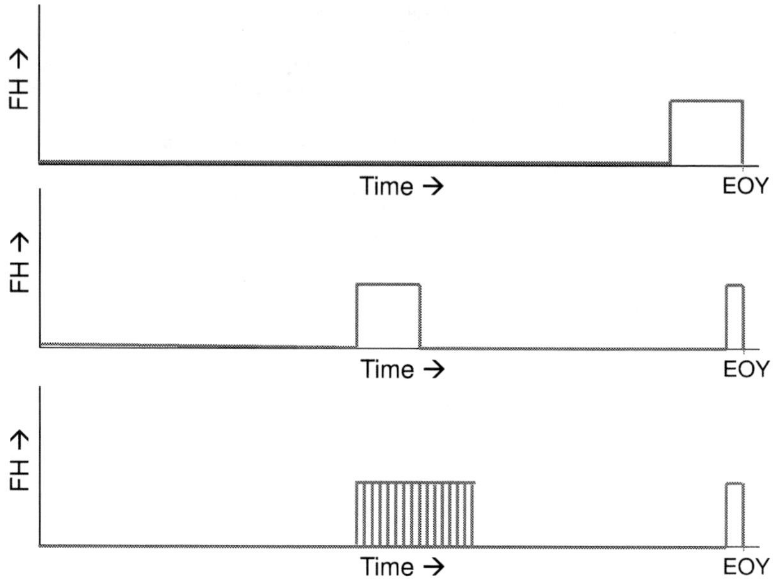

AIChE Spring Meeting
April 3, 2012

Operations Optimization – modeling challenges

Modeling challenges:

- Number of time periods
- Number of time periods with integers
- Complex electrical market
- Change of state penalties
- Cost of imported power
- Annual Fired Heater constraint

AIChE Spring Meeting
April 3, 2012

Operations Optimization – modeling challenges

Number of time periods

- Hourly time slices: 8760 periods
- Could do time lumping for more distant periods

Operations Optimization – modeling challenges

Number of time periods with integers

- Integers are required to model
  - Operating ranges
  - start up/shut down events
- Integers are computationally demanding
- Be judicious in use of integers: limit it to short term that includes the nomination time horizon

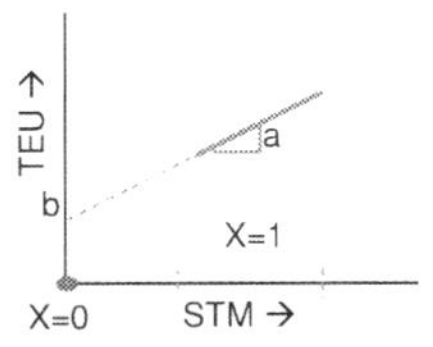

TEU = a * STM + b * X

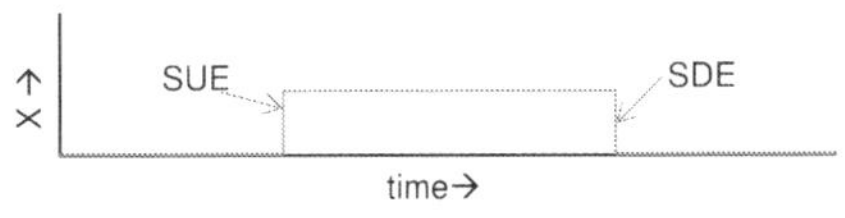

Operations Optimization – modeling challenges

Complex electrical market

- Daily, multi-round auction process
- Time dependent price
- Deviation penalties after initial nomination (smaller)
- Deviation penalties after final nomination (larger)

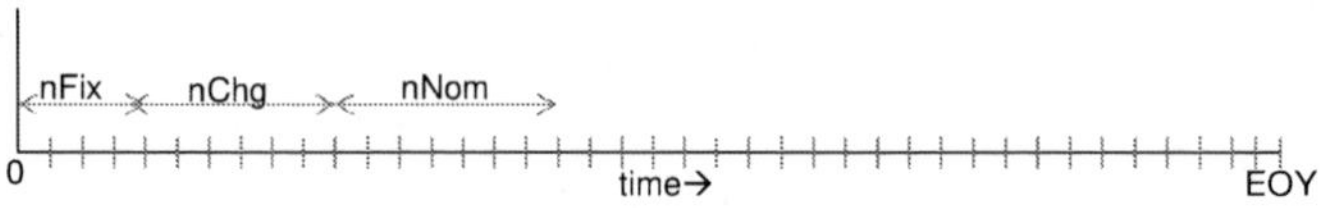

$$= \text{MAX} \sum^{EOY} c\_eee_t * EEE_t - \sum^{nFix} p\_dev_L * \Delta EEE_t - \sum^{nChg} p\_dev_S * \Delta EEE_t \ldots$$

$$\Delta EEE_t = |\ EEE_t - nom_t\ |$$

Operations Optimization - modeling challenges

Change of state penalties

- For CT, FH and PB operations
- Penalize start up and shut down events

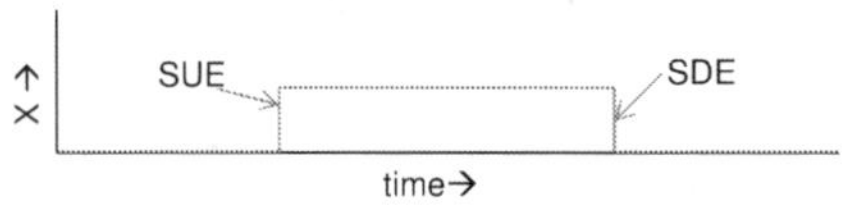

$$= \text{MAX} \ldots - \sum^{EOY} p\_sue * (CT\_SUE_t + FH\_SUE_t + PB\_SUE_t)$$
$$- p\_sde * (CT\_SDE_t + FH\_SDE_t + PB\_SDE_t) \ldots$$

- Add transition energy use and delays, if necessary

Operations Optimization - modeling challenges

Cost of imported power

- Energy cost
- Demand cost
- Excess demand penalty

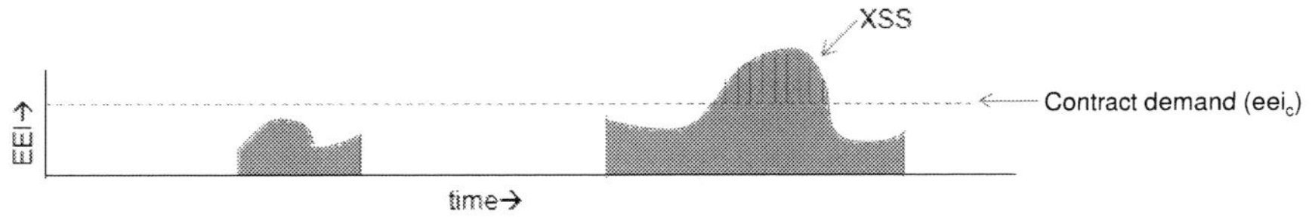

$$= \text{MAX} \ldots - \sum^{EOY} (c\_eei_t * EEI_t + c\_eed * eei_c + p\_eed * XSS_t \ldots$$

Operations Optimization - modeling challenges

Fired Heater usage constraint

- Limits total TEG as a % of total EEG (fh_{max})

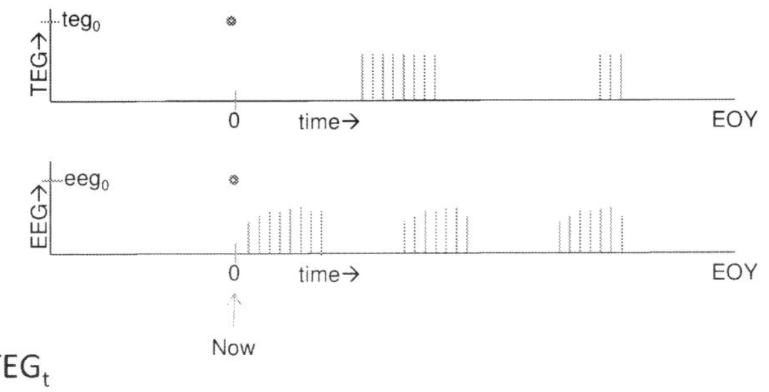

$$TEG = teg_0 + \sum_0^{EOY} TEG_t$$

$$EEG = eeg_0 + \sum_0^{EOY} EEG_t$$

$$TEG < fh_{max} / 100 * EEG$$

Operations Optimization - challenges : summary

Realistic models can be developed by addressing each of the modeling challenges.

Expect the model to be large;

- 10^3 time periods
- 10^5 continuous variables
- 10^5 constraints
- 10^2 integer variables

Target execution time < 10 minutes

Conclusions

Operation of concentrating solar power plants offers unique flexibilities and opportunities for significant profit improvement.

Large scale, multi-period, scheduling optimizer would be an essential tool to fully exploit these flexibilities.

Work is in progress towards design and implementation of operations optimizers for the three Torresol CSPs ... expected completion within the next 12 months.

Q & A

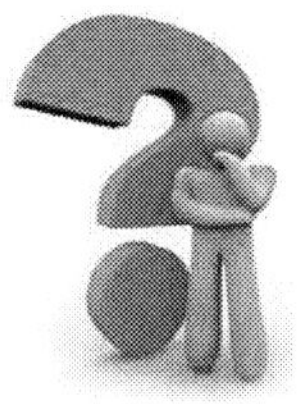

AIChE Spring Meeting
April 3, 2012

(91b) Numerical Simulations and Experiments on a Photovoltaic Solar Wall

Himanshu Dehra, 1-140 Avenue Windsor, Lachine, QC, Canada

1. Numerical SimulationsA two dimensional thermal network model is proposed to predict the temperature distribution in the photovoltaic solar wall section of an outdoor test room facility at Concordia University, Montr"al, Qu"bec, Canada. Steady-state thermal network nodal equations are developed for a photovoltaic solar wall section by neglecting thermal storage in double-glass photovoltaic module and well-insulated back panel. The enclosure analysis is assumed for obtaining radiative conductances to ascertain the effect of radiative heat exchange in between the distant composite surface nodes. One dimensional solution obtained from the heat balance model eliminates the non-linear variation of temperature dependent boundary conditions applied to thermal network. The proposed model was able to predict the temperature distribution in the test section for a range of measured climatic data under the strong influence of conjugate heat exchange.

2. Experiments The preliminary field experiments were conducted on ascertaining electrical and thermal characteristics of a pair of glass coated photovoltaic (PV) modules. Both indoor and outdoor experiments were conducted on glass coated PV modules. A pre-fabricated outdoor room was also used for conducting some of the outdoor experiments on glass coated PV modules installed on a wooden frame. The experiments on obtaining electrical characteristics of PV modules included were indoor solar simulator tests conducted under standard operating conditions, outdoor tests conducted with and without a wooden frame under available climatic conditions. Both indoor and outdoor experiments were conducted for obtaining current-voltage and current-power characteristics for a pair of PV modules. The current-voltage measurements were also obtained for determining electric power output with a series electrical circuit connection of a pair of vertically inclined PV modules installed on a wooden frame. The experiments on determining thermal characteristics included were outdoor tests conducted for obtaining temperatures, air mass velocities, sensible heat, thermal storage and thermal time constants of PV modules with and without wooden frame. After conducting a series of field experiments, preliminary results have concluded that: i) glass coated PV module installed on a wooden frame has more heat generating capacity in comparison to a window glass or a stand alone PV module; ii) the generation of electric power is affected with operation of vertically inclined PV modules; and iii) electrical and thermal characteristics are function of both ambient air temperature and thermal storage losses.

(91c) DFT Study on Hydrogen Desorption From NaMgH3 (001) Surface

Fernando A. Soto, Louisiana Tech University, Ruston, LA and Daniela S. Mainardi, Chemical engineering, Louisiana Tech University, Ruston, LA

DFT Study on Hydrogen Desorption from $NaMgH_3$(001) Surface

Authors: Fernando Soto[1], Daniela S. Mainardi[2]

[1]Institute for Micro-manufacturing, Louisiana Tech University

[2]Institute for Micro-manufacturing, Chemical Engineering, Louisiana Tech University

$NaMgH_3$ is the only magnesium based ternary hydride with Na and it has attracted the attention as a possible candidate for hydrogen storage material because it has high gravimetric hydrogen density (6 %) and high volumetric hydrogen density (88 kg/m_3). Hydrogen disassociation in this material occurs at atmospheric pressure and almost 670 K, which is well above the operable range for on board hydrogen storage applications and a major impediment for practical applications. In this study, aiming to enhance the dehydrogenation process on transition metal modified- $NaMgH_3$,(001) surface has been investigated by using density functional theory calculations. For the surface calculations, three types of models can be identified: Hydrogen desorption from the surface of the $NaMgH_3$ (H_{surf}), within the $NaMgH_3$(H_{bulk}), and at the bulk/surface interface ($H_{bulk/surf}$).

We propose that Hydrogen desorption process is facilitated in$NaMgH_3$by doping the material with Ti at the surface site and Zn as a co-dopant at the bulk site. The dehydrogenation energy of the co-doped surface structure is low, in comparison to the pristine model, Furthermore, when doped with transition metals from the 3d and 4d block elements, the dissociation barrier can also be lowered, in some cases with better results than Ti and Zn. Carbon doping was found to be the most promising, based on the considerations of the activation barrier, and dehydrogenation energy cost. Our results predict that these typesof model structures are potential useful materials for hydrogen storage application.

(91d) The Reaction Pathways of Glycerin in Supercritical Water

Jared S. Bouquet[1], Ryan E. Tschannen[2], Aaron C. Gonzales[2]and Sunggyu Lee[2], (1)Department of Chemical & Biological Engineering, Missouri University of Science and Technology, Rolla, MO, (2)Chemical and Biomolecular Engineering, Ohio University

Glycerin, a byproduct of biodiesel production, is a cheap and widely available feedstock that may be used for hydrogen production in energy application. One method that can be used to produce hydrogen from glycerin is using supercritical water as a reaction medium. The process is non-catalytic and can handle a wide range of impurities. The primary gaseous products of glycerin in supercritical water are hydrogen, carbon monoxide, methane, carbon dioxide, and ethane. In order to optimize this process, a thorough understanding of the reaction pathways of glycerin in supercritical water is needed. The different mechanistic pathways of glycerin in supercritical water were examined, including decomposition, recombination, pyrolysis, and reformation. Reactions occurring between product gases, such as the water gas shift reaction and methanation, were also examined. An experimental study was also performed in order to evaluate the accuracy of the different reaction pathways of glycerin using a 0.1L Haynes® Alloy 282 reactor system, and particular attention was given to the role of supercritical water in the reaction chemistry. Experiments were done at a variety of temperatures, space times, and water-to-glycerin molar ratios to gain a comprehensive understanding of the reaction pathways of glycerin.

(91e) Glycerol to Oxygenates and Chemicals in Supercritical Fluids

Maoqi Feng, Chemical Engineering Dept., Div. of Chemistry and Chemical Engineering, Southwest Research Institute, San Antonio, TX and Chee Kai Tan, Div. of Chemistry and Chemical Engineering, Southwest Research Institute, San Antonio, TX

Glycerol (glycerin) is a byproduct from biodiesel production via transesterification reaction from renewable vegetable oils and animal fats. With the increase of biodiesel production in the world, the glycerol supply to the market is beyond the need for chemical derivatives. The economics of biodiesel production will be improved if value-added products such as fuels or oxygenated chemicals can be produced from glycerol. This paper will address supercritical process conversion of glycerol to higher value-added chemicals and fuel components under catalytic conditions.

(91f) Autothermal Reforming of Glycerol in a Dual Layer Monolith Catalyst

Yujia Liu, Dept of Chemical Engineering and Material Science, Stevens Institute of Technology, Hoboken, NJ and Adeniyi Lawal, Dept. of Chemical Engineering and Materials Science, New Jersey Center for MicroChemical Systems, Stevens Institute of Technology, Hoboken, NJ; Chemical Engineering and Materials Science Department, Stevens Institute of Technology, Hoboken, NJ

Autothermal Reforming of Glycerol in a Dual Layer Monolith Catalyst

Yujia Liu and Adeniyi Lawal

New Jersey Center for Microchemical Systems, Dept. of Chemical Engineering and Materials Science, Stevens Institute of Technology, 1 Castle Point on Hudson, Hoboken NJ 07030

Abstract

Glycerol is a by-product during biodiesel production via transesterification of triglycerides. The conversion of glycerol to synthesis gas and then to methanol represents an important niche for glycerol processing that could relieve small biodiesel producers from high glycerol disposal cost while reducing their dependence on methanol from fossil fuels. 2O/C) ratio, oxygen to carbon (O_2/C) ratio and gas hourly space velocity (GHSV). The optimum operating conditions to produce desired conversion of glycerol, yields of H_2, and H_2/CO ratio with minimal coke formation were also determined.

(106a) Synthesis of PVA Based Membranes Modified by Benzimidazole Group for Direct Methanol Fuel Cells

Pratima Gajbhiye, Chemical Engineering, Indian Institute of Technology, Kanpur, State of Uttar Pradesh, India, Jayant K. Singh, Chemical Engineering, Indian Institute of Technology, Kanpur, Ka, India and Anil Kumar, Department of Chemical Engineering, Indian Institute of Technology Kanpur, Kanpur, India

Abstract

In recent years, there has been a great effort to look for alternate membrane material and Polybenzimidazole (PBI) membranes have been synthesized where the conducting moiety of this polymer is benzimidazolium groups. The monomers used to synthesize these polymers are bis conjugated amines and diacids are expensive. In this work, we have synthesized grafted polymers of PVA as membrane materials where the grafts are either Poly Acrylic Acid, Poly maleic anhydride (PMA) or Poly Methacrylic acid (PMAA) and these materials were denoted by PVA-g-PAA, PVA-g-PMA, and PVA-g-PMAA. We then synthesized C1, C3 & C4 compounds (formula given at the bottom of Table S1) having bis conjugated amines whereas compound C2 having only one conjugated amine group. One of the steps in synthesizing C1 compound is nitration of phenyl ring, where there is a possibility of formation of two isomers out of which only one of them has strictly bis conjugated amines and can be polymerized with Isophthalic acid giving the formation of PBI. The other byproduct formed is an impurity and is minimized by carrying out the nitration step at low temperature but this cannot be completely removed from the product which is a mixture. Table S1 presents the summary of our experimental results on characterization of membrane, In the C1 to C4 compounds and in some cases (membranes 9, 10, 11), Sulfocuccinic acid were added to the casting polymer solution as additional crosslinking agent in order to achieve higher conductivity. We added the polymer, thus formed, to the casting solution, we first compared the results of the Fuel cell formed by adding C1 (C3 and C4) compound alone and heating at 180°C for some time. Table S1 has two sets of results, one on characterization (columns 1, 2, 3,4,5,6, and 7) of membranes while the other on Fuel cell (columns 8 to 11). For the latter columns 10 and 11 gave results on characterization on Fuel cell while columns 8 and 9 gave results when the Fuel cell was run on control mode for higher efficiency.

Experiments on membrane pore sizes are given in column 1 which lies between 8.88 to 9.52nm and they all are ultra filtration type. The values of the pore size are mostly governed by the nature of the graft chains, e.g. membranes 1,3,4,7 and 12 consist of PAA grafts and were found to have similar pore sizes. Similarly membranes 2,5,8,9 were of the similar pore size because the PVA have the same grafts of PMA. The next important thing is to determine the wall charges of these membranes and they were estimated by experimentally determining the rejection of Chromic (VI) acid solution through these membranes. Subsequently, a space charge model was proposed for the flow of the solution through their pores and a semi analytical solution was derived for the overall problem of separation of chromic acid solution. We observed that the high surface charge was because of the benzimidazolium groups present on the surface and they ranged from minimum of 465 to maximum of 525mV. The thermal degradation (TGA) experiments were carried out which showed that, through the modification of the polymer by benzimidazole groups, lead to stable membrane and the degradation temperature varies from 150°C for 10% weight loss and for approximately 80% weight reduction, a minimum of 300°C to 700°C.

The twelve membranes reported in Table S1 were first characterized completely in terms of pore size, wall charge, conductivity, IEC and thermal degradation. In fuel cell applications, higher temperatures give large current density and it is observed that some of the membranes (rows 1, 2, 5, 6, 7, 8 and 12) could easily be operated upto 150°C without the membrane degrading. However in our case, we insisted that the membrane of direct methanol fuel cell were operated at 90°C for which most of the fuel cell results for Nafion 117 are available. The catalysts used on the anode and cathode side were Pt-Ru and Pt loaded on carbon respectively, which were synthesized by following procedure given in literature. We find that in each case, the power density and the current density were higher and for some of our results (membranes 1,6,12) were atleast two to three times higher than that of Nafion 117 as seen in column 10 and 11.

It was recognized that the membrane material has hydrophilic PVA matrix and hydrophobic benzimidazolium groups at the surface formed by reaction of the C1 to C4 compounds with grafts (see Fig.S-1). When the reaction occurs (confirmed by IR and NMR), the hydrophobic benzimidazolium are formed which tends to separate as confirmed by the SEM results. In view of this we assumed that the pores of the membrane would have a thin hydrophobic layer with positive charge walls. This is confirmed by the contact angle measurements where the minimum contact angle is 51° and varied upto 89°. Even though most of the polymer matrix is PVA with its grafts, when the ions move through the pores (giving rise to current in Fuel cell), the pore walls get highly charged (as measured by column 2 of Table S1). Some of the membranes have Sulfo succinic acid SSA as part of the matrix (e.g. the membrane 9 to 11) which would tend to give sulfonate groups inside the PVA matrix as well as at the phase boundaries, this way giving rise to bipolar nature of the membrane (benzimidazolium group being strictly cationic in nature). In addition to this, some of the compounds such as C1 and C2 compounds have bonded sulfonate groups which would give the positive and negative charge separated by the molecular distance, once again producing a bipolar nature to the membrane. In remainder of the cases the charges would be induced on the phase boundaries as seen in Fig.S-1. The conductivity observed in our membranes is very high and could be attributed to the special bipolar structure of our polymers.

Table S1. Fuel cell and Membrane Characterization

Compound	Membrane	1	2	3	4	5	6	7	8	9	10	11

			MWCO (pore size, nm)	Wall Charge	Conductivity (S/cm)	IEC	Thermal Degradation temp (°C) 10% - 80%		Water Uptake	Contact Angle	Average Voltage Gain (V)	Average time cycle (min)	Power Density (mW/cm)	Current Density (mA/cm)
C1	1	PVA-g-PAA-C1	9.1	488	0.04	2.7	199.732	268.70	7.5	72	0.523	55.75	50.47	74.88
	2	PVA-g-PMA-C1	9.35	465	1.23×10^{-2}	1.8	142.659	447.6	10.691	74.917	0.568	42.896	0.305	3.81
	3	PVA-g-PAA-modified	9.52	488	1.51×10^{-2}	2.12	115.123	706.53 (28.93%)	13.665	75.581	0.778	69.85	37.37	64.44
C2	4	PVA-g-PAA-C2	9.52	525	6.59×10^{-2}	2.38	119.558	691.93	13.4	51.2	0.704	39.83	23.0020	50.55
	5	PVA-g-PMA-C2	9.36	519	1.04×10^{-2}	2.17	181.59	562.348	14.9	50.24	0.77	161.93	34.72	62.11
	6	PVA-g-PMAA-C2	8.99	482	1.16×10^{-2}	2.18	201.055	472.568	13.8	51.8	0.6066	147.81	76.72	92.33
C3	7	PVA-g-PAA-C3	9.0	512	2.51×10^{-2}	2.23	204.908	453.03	8.624	70.368	0.6883	82.55	36.397	75
	8	PVA-g-PMA-C3	9.5	510	3.51×10^{-2}	2.27	166.819	505.293	13.513	62.944	0.726	37.557	36.11	64.381
	9	PVA-g-PAA-SSA-C3	9.36	497	1.71×10^{-2}	2.51	154.914	813.769 (24.32%)	10.5	54.8	0.8096	99.423	23.45	57.778
	10	PVA-g-PMA-SSA-C3	9.83	480	2.28×10^{-2}	2.46	115.12	428.713	8.5	76.2	0.663	224.22	28.561	58.358
	11	PVA-g-PMAA-SSA-C3	8.88	525	3.66×10^{-2}	1.85	139.772	681.715 (30.78%)	7.2	89.2	0.77	18.36	20.832	48.11
C4	12	PVA-g-PAA-C4	9.5	495	2.56×10^{-2}	1.62	189.87	451.86	8.927	59.132	0.704	74.19	41.1	71
	13	Nafion 117			0.1								21.01	48.33

Compound C1, [Bis-3-Amino-4-{3-(triethylammoniumsulfonato) phenylamino}-phenyl sulfone] hydrochloride

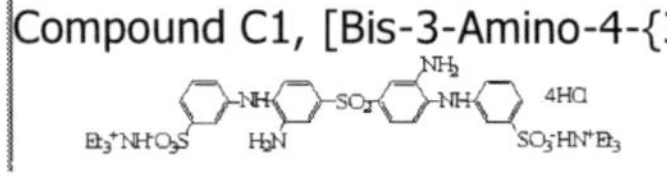

Compound C2, 3-Amino -4-[3-(triethyl ammonium sulfonato) phenyl amino] phenylene hydrochloride 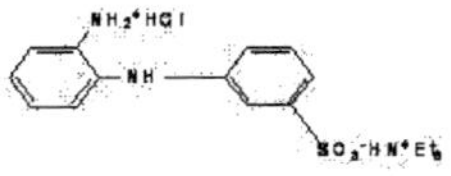
Compound C3, Bis -4 [1, 2 diamino phenyl]-phenyl sulfone) hydrochloride 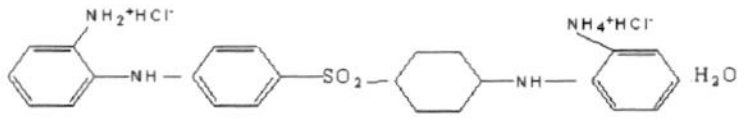
Compound C4, 1, 4 [1, 2 diamino phenyl] Phenylene hydrochloride 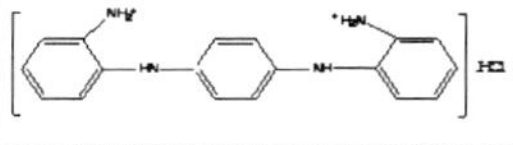

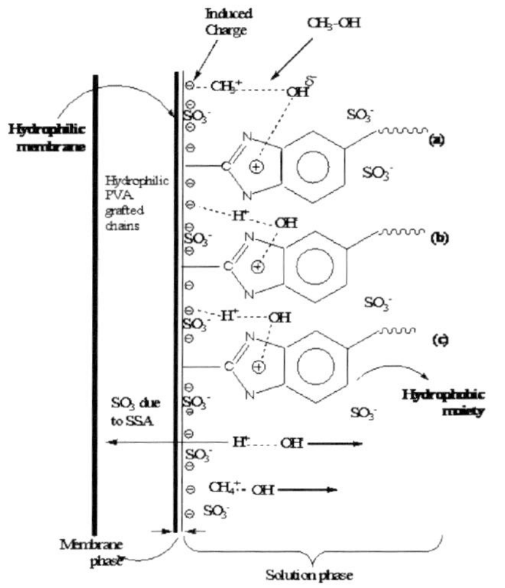

Figure S-1. The Bipolar nature of the membrane surface

(106b) Operation and Controls of Operating Direct Methanol Fuel Cells with PVA Based Membranes Modified by Benzimidazolium Groups

Pratima Gajbhiye, Chemical Engineering, Indian Institute of Technology, Kanpur, State of Uttar Pradesh, India, Pranjal Joshi, Chemical Engg, IIT Kanpur, Kanpur, India, Jayant K. Singh, Chemical Engineering, Indian Institute of Technology, Kanpur, Ka, India and A. Kumar, Department of Chemical Engineering, Indian Institute of Technology Kanpur, Kanpur, India

Operation and Control of Direct Methanol Fuel cell with PVA based membranes modified with benzimidazolium groups

Pratima Gajbhiye, Pranjal Joshi, Jayant K.Singh, Anil Kumar

Department of Chemical Engineering, Indian Institute of Technology, Kanpur 208016, India

Abstract

In recent years, in an effort to find alternate membrane material (other than Nafion 117), Polybenzimidazole (PBI) membranes have been synthesized where the conducting moiety of this polymer is benzimidazolium groups. The monomers used to synthesize these polymers are bis conjugated amines and diacids are expensive. In this work, we have synthesized grafted polymers of PVA as membrane materials where the grafts are either Poly Acrylic Acid, Poly maleic anhydride (PMA) or Poly Methacrylic acid (PMAA) and these materials were denoted by PVA-g-PAA, PVA-g-PMA, and PVA-g-PMAA. We then synthesized C1, C3 & C4 compounds (formula given at the bottom of Table S1) having bis conjugated amines whereas compound C2 having only one conjugated amine group. One of the steps in synthesizing C1 compound is nitration of phenyl ring, where there is a possibility of formation of two isomers out of which only one of them has strictly bis conjugated amines and can be polymerized with Isophthalic acid giving the formation of PBI. The other byproduct formed is an impurity and is minimized by carrying out the nitration step at low temperature but this cannot be completely removed from the product which is a mixture. Table S1 presents the summary of our experimental results on characterization of membrane, In the C1 to C4 compounds and in some cases (membranes 9, 10, 11), Sulfocuccinic acid were added to the casting polymer solution as additional crosslinking agent in order to achieve higher conductivity. We added the polymer, thus formed, to the casting solution, we first compared the results of the Fuel cell formed by adding C1 (C3 and C4) compound alone and heating at 180°C for some time. Table S1 has two sets of results, one on characterization (columns 1, 2, 3,4,5,6, and 7) of membranes while the other on Fuel cell (columns 8 to 11). For the latter columns 10 and 11 gave results on characterization on Fuel cell while columns 8 and 9 gave results when the Fuel cell was run on control mode for higher efficiency.

Experiments on membrane pore sizes are given in column 1 which lies between 8.88 to 9.52nm and they all are ultra filtration type. The values of the pore size are mostly governed by the nature of the graft chains, e.g. membranes 1,3,4,7 and 12 consist of PAA grafts and were found to have similar pore sizes. Similarly membranes 2,5,8,9 were of the similar pore size because the PVA have the same grafts of PMA. The next important thing is to determine the wall charges of these membranes and they were estimated by experimentally determining the rejection of Chromic (VI) acid solution through these membranes. Subsequently, a space charge model was proposed for the flow of the solution through their pores and a semi analytical solution was derived for the overall problem of separation of chromic acid solution. We observed that the high surface charge was because of the benzimidazolium groups present on the surface and they ranged from minimum of 465 to maximum of 525mV. The thermal degradation (TGA) experiments were carried out which showed that, through the modification of the polymer by benzimidazole groups, lead to stable membrane and the degradation temperature varies from 150°C for 10% weight loss and for approximately 80% weight reduction, a minimum of 300°C to 700°C.

The twelve membranes reported in Table S1 were first characterized completely in terms of pore size, wall charge, conductivity, IEC and thermal degradation. In fuel cell applications, higher temperatures give large current density and it is observed that some of the membranes (rows 1, 2, 5, 6, 7, 8 and 12) could easily be operated upto 150°C without the membrane degrading. However in our case, we insisted that the membrane of direct methanol fuel cell were operated at 90°C for which most of the fuel cell results for Nafion 117 are available. The catalysts used on the anode and cathode side were Pt-Ru and Pt loaded on carbon respectively, which were synthesized by following procedure given in literature. We find that in each case, the power density and the current density were higher and for some of our results (membranes 1,6,12) were atleast two to three times higher than that of Nafion 117 as seen in column 10 and 11. In normal Fuel cell operations, there is decay in the cathode catalyst which is further increased by the methanol crossover. The voltage of the Fuel cell increases, the average volatage gained is reported in column 8 and it is found that the voltage gain depends on the nature of the membrane which is not the same for all of them. In column 9, we have measured the average time cycle and we found that some of the membranes (membranes 6, 7, 9, 10) gave extremely good results.

It was recognized that the membrane material has hydrophilic PVA matrix and hydrophobic benzimidazolium groups at the surface formed by reaction of the C1 to C4 compounds with grafts (see Fig.S-1). When the reaction occurs (confirmed by IR and NMR), the hydrophobic benzimidazolium are formed which tends to separate as confirmed by the SEM results. In view of this we assumed that the pores of the membrane would have a thin hydrophobic layer with positive charge walls. This is confirmed by the contact angle measurements where the minimum contact angle is 51° and varied upto 89°. Even though most of the polymer matrix is PVA with its grafts, when the ions move through the pores (giving rise to current in Fuel cell), the pore walls get highly charged (as measured by column 2 of Table S1). Some of the membranes have Sulfo succinic acid SSA as part of the matrix (e.g. the membrane 9 to 11) which would tend to give sulfonate groups inside the PVA matrix as well as at the phase boundaries, this way giving rise to bipolar nature of the membrane (benzimidazolium group being strictly cationic in nature). In addition to this, some of the compounds such as C1 and C2 compounds have bonded sulfonate groups which would give the positive and negative charge separated by the molecular distance, once again producing a bipolar nature to the membrane. In remainder of the cases the charges would be induced on the phase boundaries as seen in Fig.S-1. The conductivity observed in our membranes is very high and could be attributed to the special bipolar structure of our polymers.

Table S1. Fuel cell and Membrane Characterization

Compound	**Membrane**		**1**	**2**	**3**	**4**	**5**		**6**	**7**	**8**	**9**	**10**	**11**
			MWCO (pore size, nm)	**Wall Charge**	**Conductivity (S/cm)**	**IEC**	**Thermal Degradation temp (°C) 10% - 80%**		**Water Uptake**	**Contact Angle**	**Average Voltage Gain (V)**	**Average time cycle (min)**	**Power Density (mW/ cm)**	**Current Density (mA/ cm)**
C1	1	PVA-g-PAA-C1	9.1	488	0.04	2.7	199.732	268.70	7.5	72	0.523	55.75	50.47	74.88
	2	PVA-g-PMA-C1	9.35	465	1.23×10^{-2}	1.8	142.659	447.6	10.691	74.917	0.568	42.896	0.305	3.81
	3	PVA-g-PAA-modified	9.52	488	1.51×10^{-2}	2.12	115.123	706.53 (28.93%)	13.665	75.581	0.778	69.85	37.37	64.44
C2	4	PVA-g-PAA-C2	9.52	525	6.59×10^{-2}	2.38	119.558	691.93	13.4	51.2	0.704	39.83	23.0020	50.55
	5	PVA-g-PMA-C2	9.36	519	1.04×10^{-2}	2.17	181.59	562.348	14.9	50.24	0.77	161.93	34.72	62.11
	6	PVA-g-PMAA-C2	8.99	482	1.16×10^{-2}	2.18	201.055	472.568	13.8	51.8	0.6066	147.81	76.72	92.33
C3	7	PVA-g-PAA-C3	9.0	512	2.51×10^{-2}	2.23	204.908	453.03	8.624	70.368	0.6883	82.55	36.397	75
	8	PVA-g-PMA-C3	9.5	510	3.51×10^{-2}	2.27	166.819	505.293	13.513	62.944	0.726	37.557	36.11	64.381
	9	PVA-g-PAA-SSA-C3	9.36	497	1.71×10^{-2}	2.51	154.914	813.769 (24.32%)	10.5	54.8	0.8096	99.423	23.45	57.778
	10	PVA-g-PMA-SSA-C3	9.83	480	2.28×10^{-2}	2.46	115.12	428.713	8.5	76.2	0.663	224.22	28.561	58.358
	11	PVA-g-PMAA-	8.88	525	3.66×10^{-2}	1.85	139.772	681.715	7.2	89.2	0.77	18.36	20.832	48.11

		SSA-C3						(30.78%)						
C4	12	PVA-g-PAA-C4	9.5	495	2.56 x10^{-2}	1.62	189.87	451.86	8.927	59.132	0.704	74.19	41.1	71
	13	Nafion 117			0.1								21.01	48.33

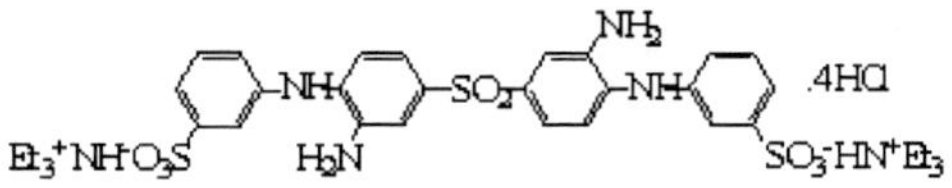

Compound C1, [Bis-3-Amino-4-{3-(triethylammoniumsulfonato) phenylamino}-phenyl sulfone] hydrochloride

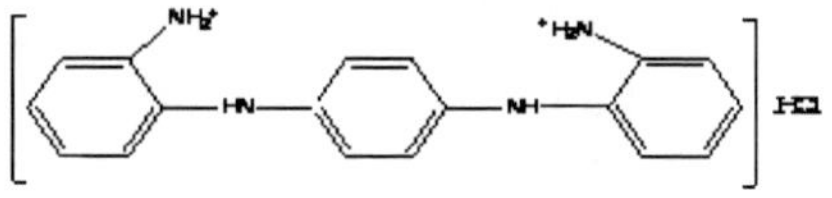

Compound C3, Bis -4 [1, 2 diamino phenyl]-phenyl sulfone) hydrochloride

Compound C2, 3-Amino -4-[3-(triethyl ammonium sulfonato) phenyl amino] phenylene hydrochloride

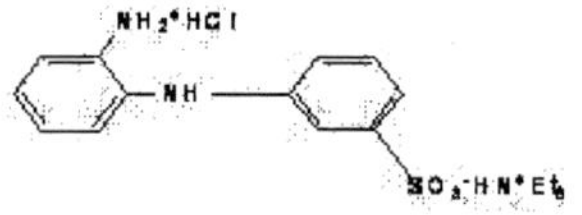

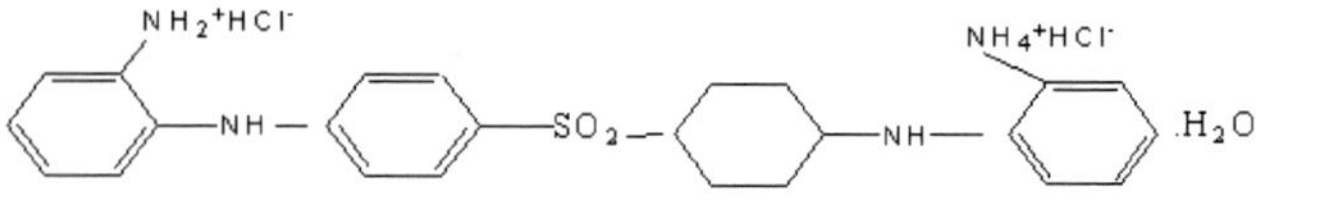

Compound C4, 1, 4 [1, 2 diamino phenyl] Phenylene hydrochloride

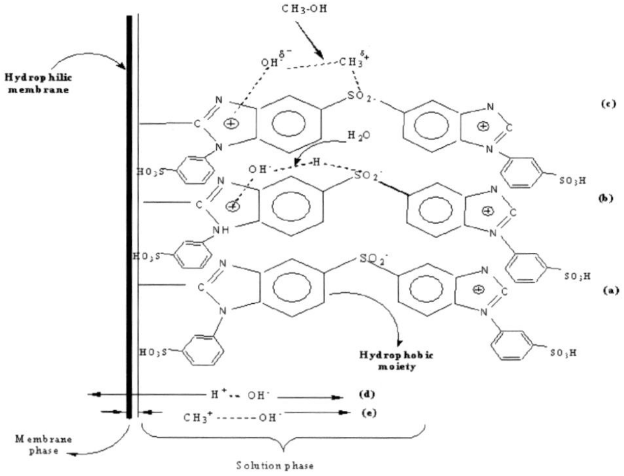

Figure S-1. The bipolar nature of PVA-g-PAA-BASPAPS membrane which reduces the methanol crossover due to the molecular dissociation. In (a), the imidazolium group is shown to separate out from the PVA moiety of the chain into a separate phase. In (b), the water is shown to complex with the SO and the imidazolium group. As a result of this, interaction in (d) and (e), the water and the methanol is shown to split into anion and cation and the anion permeates through the membrane giving bipolar effect

(106e) Numerical Investigation in Heat Conduction Effect on Heat Exchange for Mini Cross-Flow Heat Exchanger (MCHE)

Hongyue Duan[1], Wei Mu[1], **Xigang Yuan**[1] and Lingai Luo[2], (1)State Key Laboratory of Chemical Engineering, Tianjin University, Tianjin, China, (2)LOCIE Chambery, Savoie University, Savoie, France

In a traditional heat exchanger, the effect of heat transfer within the wall material driven by the temperature gradient in axial direction of the fluid flow is usually very small because the area of the cross section of the tube wall is much smaller than that of the tube for heat exchange. However, in a mini cross-flow heat exchanger(MCHE), where the heat conducting medium is typically a cubic-shaped block of metal and the hot and cold streams are put into heat exchange via two sets of channels in the block, the heat conduction in the wall material in the direction parallel to the fluid flow may have a significant effect on the performance of the heat exchanger.

In this paper, a 2-channel-layer MCHE was simulated and a periodic boundary condition was applied so that the results could be applicable to entire MCHE. The proposed model and the mesh grid size were proved by the good agreement of the simulated overall heat transfer coefficient with that found in the experimental reported in the literature[1]. Then the effect of heat conduction on heat transfer performance in a MCHE was studied numerically. The simulation results suggest that the performance of the MCHE can be improved by increasing the wall thickness between the adjoining channels in the same layer or by decreasing the wall thickness between the adjacent layers. The heat conduction effect can also be weakened by optimizing the material conductivity or by increasing the diameter of the channels. New correlations were developed based on the simulated results to predict the value of the overall heat transfer coefficient as a function of the operating parameter such as Re in the channel and the design parameters of the MCHE such as the diameter of the channel, wall thickness and thermal conductivity.

[1]Luo L A, Fan Y L, Zhang W W ***et al.***Integration of constructal distributors to a mini crossflow heat exchanger and their assembly configuration optimization[J]. Chemical Engineering Science, 2007, 62 (13): 3605-3619.

(106g) The Effects of Out Gassing Through Cylindrical Walls on the Transport Coefficients

Luma. H. Yaseen[1],**Nada B. Nakkash**[2]and M. N. Latif[1], (1)Al-Nahrain University, Baghdad, Iraq, (2)Chemical Engineering, Al-Nahrain University, Baghdad, Iraq

Abstract

This work concerns with studying the influence of blowing on friction factor and mass transfer coefficient when the flow is normal to a tube bank with outgassing porous cylinders. The friction factor and the mass transfer coefficients are measured for a range of Reynold's number and blowing rate and expressed in terms of various dimensionless groups.

The test rig consists of a rectangular duct (20.3 × 11.3 cm), five rows of three porous tubes in staggered arrangements fixed across the duct with 1.5pitches (both longitudinal and transversal). The outside and the inside diameters of tubes were 25.4 and 21.4 mm, respectively.

The friction factor measurements was studied for Reynold's number range of

$3000 < Re < 17,000$and blowing rate up to 4×10^{-3}. The results show that the blowing intothe boundary layer for the cylinders increases the overall friction factor from 10%to 50% depending on the blowing rate. The mass transfer coefficient was measured for Reynold's number range of $5000 < Re < 16,000$ and blowing rate of $10^{-5} < m/G_{\infty} < 10^{-3}$.

A new definition of the blowing rate was considered and expressed by the mass transfer driving force. Two successful formulas were obtained to account the outgassing effect on the mass transfer coefficient. The mass transfer coefficient for the outgassing boundary layer was obtained first by determining the non-blowing coefficient and multiplying it with the correction factor .

The first model was modified to account the density variation associated with the boundary layer change:

$$St_m Sc^{2/3} = \frac{0.51\, Re^{-0.5}}{(1+\beta_m)^{4.7}} \left(\frac{\rho_s}{\rho_\infty}\right)^{0.4}$$

With the coefficient of determination R2 = 0.93.

The second model depends on the flat plate correction factor:

$$St_m Sc^{2/3} = 0.5\, Re^{-0.5} \left(\frac{Ln(1+\beta_m)}{\beta_m}\right)^{9.94}$$

With the coefficient of determination R2 = 0.92.

These two models are found to be in agreement with the previous literature and a reduction of about 80% in the mass transfer coefficient has been obtained due to the outgassing.

(106h) Down the Hubbert Curve

Ian Sutton, Sutton Technical Books, Houston, TX

Down the Hubbert Curve

In March 1956 M. King Hubbert, who was at that time Chief Geologist with the Shell Development Company published a seminal paper with the title "Nuclear Energy and the Fossil Fuels" (Hubbert 1956). In this lengthy and authoritative paper, which was presented before the American Petroleum Institute in San Antonio, Texas, Dr. Hubbert showed how key fossil fuel resources in the United States, particularly coal and oil, were of finite size and that their output would reach a maximum, and then decline. Dr. King's forecasts for the production of oil in the United States were accurate; he predicted that oil production would peak in the early 1970s, and that is what has happened. (The widely-used term "Peak Oil" is frequently accredited to Dr. Colin Campbell (2001)).

In the fifty years since the publication of Dr. King's paper great advances have been made in finding and extracting oil from what were considered to be marginal fields. But Dr. King's overall conclusion is still valid. Production of oil in the United States remains below the 1970's peak. Although Dr. Hubbert's analysis was confined to the United States, subsequent analysts applied his insights to other parts of the world. For example, the book ***Twilight in the Desert: The Coming Saudi Oil Shock and the World Economy***(Simmons 2005) by the late Matthew Simmons achieved wide recognition by showing that the concept of peak oil most likely applies to reserves in Saudi Arabia, and other nations such as Iran and Kuwait.

The purpose of this paper is to examine how oil production and consumption have changed in the last five years, and to examine whether or not the concept of Peak Oil remains valid. Key to this discussion is an understanding of the concept of Energy Returned on Energy Invested (EROEI). It is concluded that the concept of Peak Oil, as first postulated by Dr. King, remains valid and current, and the production of oil peaked in the period 2005-2011 and is now set to move down the right hand side of the Hubbert Curve.

The title of this session is ***Challenges in Instituting Energy Savings Project in the Present Market Conditions***. If the production of oil and has indeed already passed a peak then the challenge is not to do with "instituting" energy savings; those savings are going to occur whether we like it or not. The challenge will be to determine how the process industries can restructure themselves given that the energy resources available (both as direct energy and as petrochemical feedstock) will be in decline and that there will be intense demand for those same resources from other areas of the economy.

(106j) Statistical Decision Making for In-Service Inspection of Underground Waste Storage Tanks

Emily Mitchell, Emory University, Atlanta, GA and **Stephen P. Harris**, Savannah River National Laboratory, Aiken, SC

Pitting is a localized form of corrosion that occurs on passivated metal surfaces exposed to aggressive chemical environments. The pitting process involves the breakdown of the passive oxide film and subsequent anodic dissolution of the metal at a local site, while the remainder of the surface remains protected by the film.

Deep pits may result in lowering the maximum level of waste that made be stored in the tank and eventually lead to loss of confinement. Pit depths, established by ultrasonic testing, are subject to variables that cause uncertainty, including the repeatability of metric location, time between measurements, different transducers, and different inspectors.

The generalized extreme value (GEV) distribution is fit to the data gathered from ultrasonic testing and statistical projections are made for the full tank wall interior. A comparison is made with results from other well known statistical distributions. The estimated parameters are used in statistical tolerance limits for bounding pit depth in a waste tank.

The tolerance limits give an indication of extreme values of pit depth for the current condition of the tank walls. The upper tolerance limit for pit depth will then be compared to action levels for pitting. If the upper tolerance limit for pit depth exceeds the action level, more frequent examinations or increased tank surface area inspections are conducted.

(106k) Thermal Inbreathing Requirements of Low Pressure Storage Tanks At Elevated Temperatures

Donald E. Brooks, Pressure Relief Analysis, Lloyd's Register Energy Americas, Inc., Houston, TX

ISO 28300 (API 2000) "Petroleum, petrochemical and natural gas industries - Venting of atmospheric and low pressure storage tanks" provides guidance on determining the required thermal inbreathing flow rates to protect atmospheric and low pressure storage tanks against vacuum due to weather changes (e.g. a decrease in atmospheric temperature)." For heated, un-insulated tanks with vapor space temperatures above 58.9 °C (120 "F), ISO 28300 recommends an engineering review of thermal inbreathing requirements.

This paper describes one approach to evaluating thermal inbreathing requirements for un-insulated storage tanks with vapor space temperatures exceeding 58.8 °C (120 "F)." The tank is assumed to be exposed to the cooling effects of a sudden rain. A simplified unsteady state heat transfer analysis is performed to determine the tank wall and vapor space temperatures versus time. Initial gas mass and pressure are assumed to remain constant in the tank. Gas volume in the tank is calculated as a function of vapor space temperature using the ideal gas law. The rate of volume change is calculated for each time increment. The rate of volume change goes through a maximum value. This maximum value is selected as the thermal inbreathing requirement of the tank.

The calculation methodology described in this paper can be programmed in a spreadsheet for ease of use.

Dynamic Operational Risk Assessment with Bayesian Networks

Shubharthi Barua, Xiaodan Gao, M. Sam Mannan
Mary Kay O'Connor Process Safety Center
Texas A&M University, Jack E. Brown Engineering Bldg., 3122 TAMU, Room 244,
College Station, TX 77843
shubharthi.barua@tamu.edu, xiaodan_gao08@mail.che.tamu.edu, mannan@tamu.edu

Extended Abstract Prepared for the Poster Presentation at the
8th Global Congress on Process Safety
Houston, TX
April 1 – 4, 2012

UNPUBLISHED

Dynamic Operational Risk Assessment with Bayesian Networks

Shubharthi Barua, Xiaodan Gao, M. Sam Mannan

Mary Kay O'Connor Process Safety Center
Texas A&M University, Jack E. Brown Engineering Bldg., 3122 TAMU Room 244, College Station, TX-77843
shubharthi.barua@tamu.edu, xiaodan_gao08@mail.che.tamu.edu, mannan@tamu.edu

Keywords: risk assessment, Bayesian network, dynamic Bayesian network, fault tree, event tree

Abstract

Oil/gas and petrochemical plants are complicated and dynamic in nature. Dynamic characteristics include ageing of equipment/components, season changes, stochastic processes, operator response times, inspection and testing time intervals, and timing of safety system operations, all of which are time dependent criteria that can influence dynamic processes. Conventional risk assessment methodologies can quantify dynamic changes in processes with limited capacity. Several methods are proposed for dynamic risk assessment such as C-DET, D-DET, DORA, etc., but all of them have deficiencies of some kind. Bayesian networks, a graphical inference technique, have some advantageous characteristics for risk analysis over other techniques. This research focuses on providing a framework for dynamic risk analysis based on Bayesian networks. A case study will be provided to illustrate this model.

1.0 Introduction

Oil/gas, chemical, and petrochemical process plants consist of numerous equipment and unit operations, thousands of control loops, and exhibit dynamic behavior. Different hazards and several types of risk are associated with these process facilities. For example, large process areas have the potential to stock a large amount of hazardous chemicals, which can result in the process area being highly congested. It is reported that process risk, i.e., risk in process facilities due to fires and explosions, greatly contributes to the total risk of offshore installations. Also, reactivity, toxic, and mechanical hazards have a significant effect on the total risk of these facilities. Therefore, it is very important to identify hazards, perform risk assessments, and take proper initiatives to minimize/remove hazards and risks; else a catastrophic accident may result.

From case histories, it has been observed that catastrophic accidents have a significant effect on people, environment, and society. Catastrophic accidents such as the Flixborough disaster, the Bhopal incident, and the Piper Alpha disaster caused fatalities and unbearable economic loss. In the past few years, several disastrous accidents in refineries, power plants, and offshore platforms have significantly affected people's perception, and consequently raised concern to emphasize process safety. A series of explosions in 2005 resulting from insufficient process safety and lack of risk reduction measures at the BP Texas City refinery accident resulted in 15 fatalities, 180 injuries, and $1.5 billion in losses [1]. The Kleen Energy power plant explosion in

Connecticut in 2010 is considered to be one of the worst industrial explosions in the USA [2]. In 2010, a fire and explosion resulting from a blowout at the Macondo well resulted in 11 deaths and 17 injuries. Also the continuous spill from the wellhead for 87 days had disastrous effects on the environment and wildlife surrounding the Gulf of Mexico. It has been reported that most of the accidents could be prevented or their severity could have been minimized by effective risk assessment and process safety management. But existing risk assessment methodologies cannot properly quantify various time-dependent effects. Therefore, the objective of this research is to develop a risk assessment methodology with the capability to capture variations due to different time-dependent effects.

2.0 Literature Review

N. Siu [3] summarized available methodologies for risk assessment. For example, the fault tree and event tree methods are two most common methodologies for risk assessment. But both techniques have limited ability to address different time-dependent effects such as equipment or component ageing, stochastic processes, operator response times, and inspection and testing time intervals. N. Siu [3] also described other methodology limitations while applying dynamic system analyses. Therefore, for dynamic systems, conventional techniques are not able to provide accurate results because of their inadequate ability to describe the variation of operational risks as time-dependent deviations, or the changes occurring in the process. Continuous event trees [4], discrete event trees [5], and Markov theory [6] methods have been proposed for dynamic reliability and safety analysis. But all of the proposed techniques have some deficiencies. For example, the Markov theory based model does not consider the effect of inspection on system-state transitions. X. Yang et al. [1] proposed the Dynamic Operational Risk Assessment (DORA) technique, which is able to take inspection and testing time intervals and repair time into account for evaluating operational failure.

Recently, Bayesian networks are receiving wide acceptance in reliability assessment, fault diagnosis, safety, risk, and maintenance analysis. Several authors have [7,8] compared some commonly used methods for accident analysis with Bayesian networks, and described several advantages of Bayesian networks over these methods. N. Khakzad et al. [2] illustrated parallelism between Bayesian networks and fault trees, and discussed the potential of Bayesian networks as a tool for process safety and accident analysis. H. Pasman and W. Rogers [9] described an approach to incorporate Bayesian networks into layer of protection analysis to make it more effective. These studies demonstrate the potential advantages of Bayesian networks application in risk assessment. This research provides a methodology for dynamic risk assessment based on Bayesian networks.

3.0 Methodology and Expected Result

3.01 Bayesian Networks

A Bayesian network, which is a probabilistic graphical model, describes causal influence relations among variables. A set of random variables and their conditional dependencies are represented by Bayesian networks via a directed acyclic graph (DAG). It has the ability to represent dependency among events clearly, accommodate multi mode and continuous random variables, and incorporate information i.e. generic, system specific and expert judgment to support optimum decision making.

A simple Bayesian network is shown in Figure 1. In a binary network, variables are represented by nodes and causal relationships among different nodes are shown by an arc. Conditional probability tables or defined probabilistic relations among nodes represent how one variable is linked with one or many variables. The nodes that influence another variable and have unconditional probability are called parent or root nodes. Nodes that are conditionally dependent on their direct parents are called intermediate nodes. The top node is defined as a leaf node, which is similar to the top event of a fault tree and initiating event for an event tree.

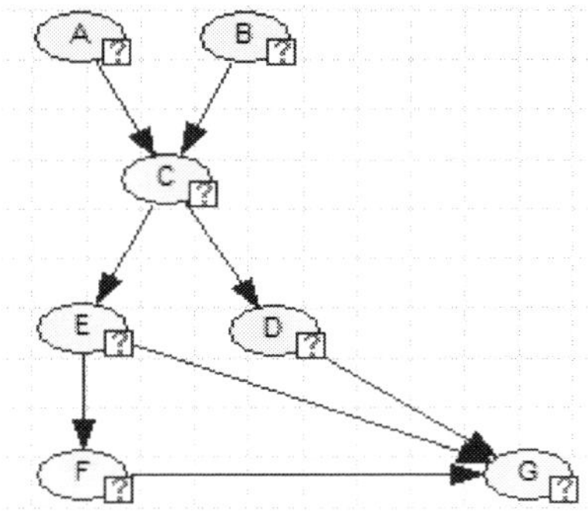

Figure 1: A simple Bayesian network

Let $N = (G, P)$ be a Bayesian network, where

$G = (V, E)$ is a directed acyclic graph; V (random variables) represents nodes; and E represents edges between pairs of nodes of DAG.

P represents probability distribution over V and $V = \{X_1, X_2, \ldots, X_n\}$ is either discrete or continuous random variables [10]. These random variables are assigned to the nodes and the edges.

Bayesian networks can be represented by the joint probability distribution P(V);

$$P(V) = \sum_{X \in V} P\{X|pa(X)\} = P(X_1, X_2, \ldots, X_N) = \sum_{i=1}^{n} P\{X_i|pa(X_i)\}$$

Here $pa(X_i)$ = parent nodes of X_i.

A main advantage of the application of Bayesian networks in risk analysis is the ability to update prior data using Bayes' theorem by incorporating new information as demonstrated by N. Khakzad et al. [2]. Also, Bayesian networks have an advantage of handling different types of uncertainty.

3.02 Dynamic Bayesian Networks

A general Bayesian network is a static model, i.e., the joint probability distribution is usually a representation of a fixed point or an interval of time [11]. A dynamic Bayesian network is a Bayesian network that has the ability to describe the evolution of joint probability distribution over time. Dean and Kanazawa [12] proposed modeling time in a discrete manner to represent the progression of time in a dynamic Bayesian network. In a dynamic Bayesian network, arcs links node/nodes from previous time slice to the node/nodes of the next time slice to represent temporal dependencies among them.

3.03 Methodology

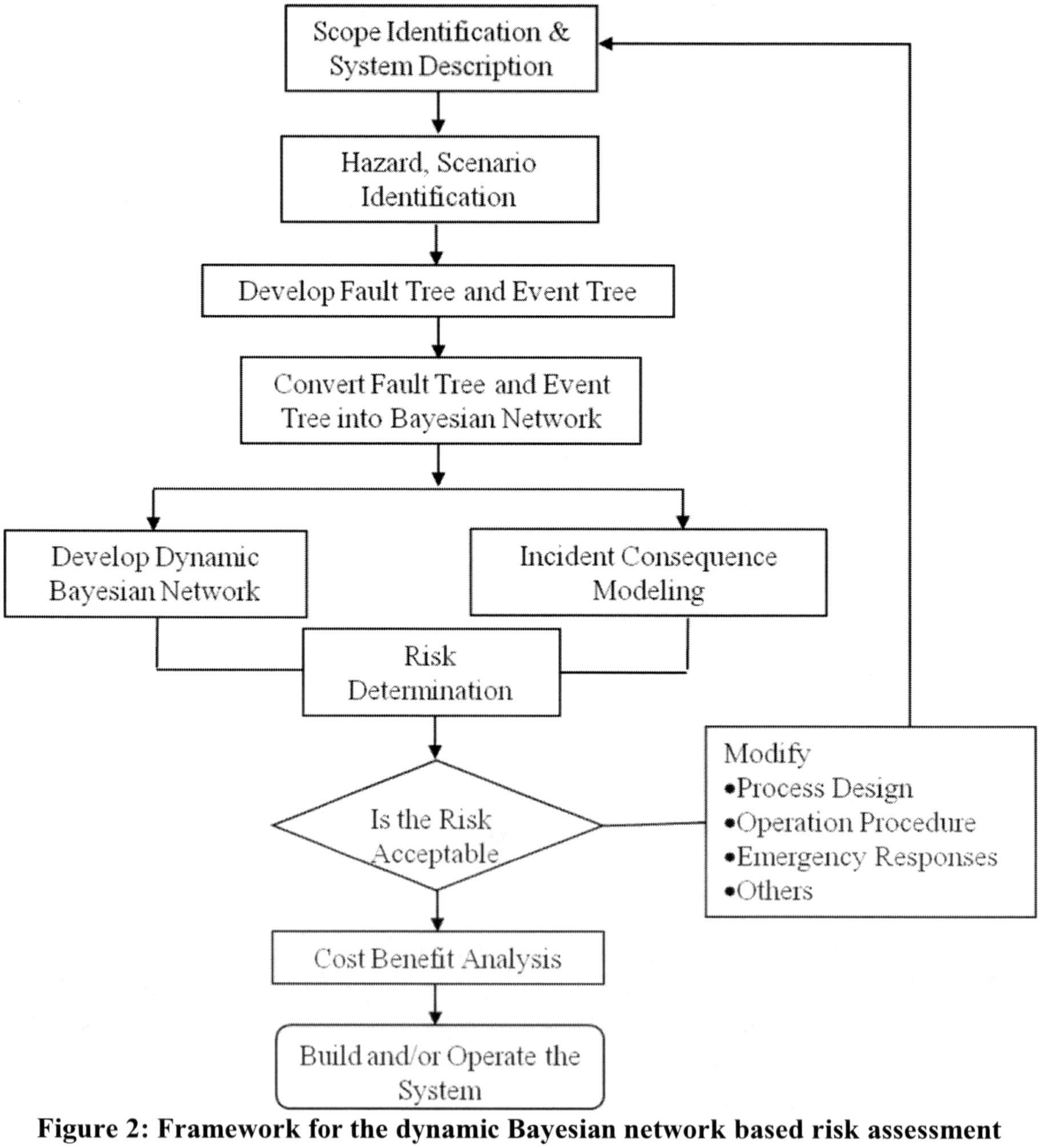

Figure 2: Framework for the dynamic Bayesian network based risk assessment

The objective of this research is to propose a dynamic risk assessment method based on dynamic Bayesian networks. According to the framework, a fault tree and event tree are developed for a system. The objective for following step is to convert the developed fault and event tree into a Bayesian network. Bobbio et al. [13] provided a methodology for conversion of fault trees into Bayesian networks. N. Khakzad et al. [2] showed equivalency between fault trees and Bayesian networks, and described the advantages of Bayesian networks for process safety analysis. Bearfield and March [14] proposed another methodology for generalizing event trees into Bayesian networks. The method has the ability to generalize simple event trees into Bayesian networks in a relatively easy manner, but the method becomes complicated for large event trees. Therefore, in the first phase, a methodology is proposed to convert fault trees and event trees into a static Bayesian network. In the second phase, a dynamic Bayesian network is developed from the static Bayesian network. The dynamic Bayesian network provides probability of a system at different time-slices for different states of the system. The probabilities are then combined with consequence modeling levels to calculate risk. GeNIe, which was developed at the Decision Systems Laboratory at the University of Pittsburgh, is used to create the model [15].

4.0 Conclusions

This article briefly describes a framework based on Bayesian network for dynamic risk assessment for the oil/gas, chemical, and petrochemical industries. This research illustrates the capability of Bayesian networks to represent all component interactions in complex systems, and to update all prior system probabilities in the presence of new information. The advantages of modeling flexibility and efficient computational mechanisms using dynamic Bayesian networks are also discussed.

4. References

[1] Yang, X. and Mannan, M.S., "The development and application of dynamic operational risk assessment in oil/gas and chemical process industry", Reliability Engineering and System Safety 95 (2010) 806-815

[2] Khakzad, N., Khan, F. and Amyotte,P., "Safety analysis in process facilities: Comparison of fault tree and Bayesian network approaches", Reliability Engineering and System Safety 96(2011) 925-932

[3] Siu,N. "Risk assessment for dynamic systems: an overview", Reliability Engineering and System Safety 43 (1994) 43-73

[4] Devooght, J. and Smidts,C. "Probabilistic reactor dynamics-I: the theory of continuous event trees", Nuclear Science and Engineering 1992; 111:229-40

[5] Amendola, A. "Accident sequence dynamic simulation versus event trees", Reliability Engineering and System Safety 1988; 22:3-25

[6] Howard, R. "Dynamic probability systems", Vols. I and II, New York, 1971

[7] Sklet, S. "Comparison of some selected methods for accident investigation", Journal of Hazardous Materials 2004; 111:29-37

[8] Nivolianitou, Z.S., Leopoulos, V.N. and Konstantinidou, M., "Comparison of techniques for accident scenario analysis in hazardous systems", Journal of Loss Prevention in the Process Industries 2004; 17"467-75

[9] Pasman, H. and Rogers, W.; "BBN, a tool to make LOPA more effective, QRA more transparent and flexible, and therefore to make safety more definable", 14th Annual Sysposium, Mary Kay O'Connor Process Safety Center, College Station, Texas, October 2011

[10] Donohue, S.K. and Dugan, J.B.; "Modeling the "Good enough to release" decision usin V&V preference structure and Bayesian belief networks", Reliability and Maintainability Symposium, 2003 Annual, 2003

[11] McNaught, K.R. and Zagorecki, A. "Using Dynamic Bayesian networks for prognostic modeling to inform maintenance decision making", IEEM 2009, Hond Kong, January 2010

[12] Dean, T. and Kanazawa, K., "A model for reasoning about persistence and causation", Computational Intelligence 1989; 5:142-150

[13] Bobbio A, Portinale L, Minichino M and Ciancamerla E., "Improving the analysis of dependable systems by mapping fault trees into Bayesian networks", Reliability Engineering and System Safety 71(3): 249-260

[14] Bearfield G and Marsh W, "Generalising event trees using Bayesian networks with a case study of train derailment", Available online at http://www.agenarisk.com/resources/white_papers/Generalising_Event_Trees.pdf , last modified: 23 February 2010, accessed: 7 January 2012

[15] Decision Systems Laboratory, University of Pittsburgh, GeNIe (Graphical Network Interface and SMILE (Structural Modeling, Inference, and Learning Enginer), Version 2.0 software, 2010. Available online at http://genie.sis.pitt.edu

(137b) Comparing the Performance of Compact Heat Exchange Reactor to CSTR for Fischer Tropsch Synthesis

Hani Gadalla[1], Zhijun Jia[1], Steven J. Vallee[1], Burtron H. Davis[2]and Dennis Sparks[2], (1)Chart Energy & Chemicals, La Crosse, WI, (2)Center for Applied Energy Research, University of Kentucky, Lexington, KY

A commercial Fischer Tropsch catalyst was studied in two Compact Heat Exchange Reactors (CHER) with different channel sizes, and in a continuous stir tank reactor (CSTR). The CHERs were constructed by Chart and Chemicals with one reactor having a 5mm channels and the other having a 6mm channels, while maintaining the same catalyst volumeThe catalyst was crushed and sieved retaining particles ranging in size from 75-125μm. Testing was performed at the University of Kentucky Center for Applied Energy Research in Lexington, KY. The CO conversion, selectivity and hydrocarbon productivity were measured for each reactor over a wide range of operating conditions, as a function of temperature, pressure, H2:CO feed ratio, and flow rate. Both reactor types showed excellent temperature control, but the performances varied at different operating conditions. The larger channel CHER had a higher conversion than the 5mm channel, potentially due to slightly higher reactor bed temperature. The two CHERs approached and even exceeded the productivity of the CSTR for liquid hydrocarbons per gram of catalyst. At higher flow rates the methane selectivity increases at a lower rate for either of the CHERs than for the CSTR.

Best Practices in the Design and Fabrication of Pilot and Demonstration Plants

David Beckman, President and CEO, Zeton Inc., Burlington, ON, Canada
Adam Whalley, Business Development Manager, Zeton Inc., Burlington, ON, Canada

Presentation ID# 246699

AIChE
2012 Spring Meeting
April 1-5, 2012
Hilton Americas-Houston and George Brown Convention Center
Houston, TX

As a specialty designer and fabricator of modular pilot plants, one of the first questions asked of Zeton in the early stages of a project is 'how can we speed up our project, and get the pilot plant delivered in less time?' The engineers asking this question are experienced professionals that understand project management. They already have a firm grasp on fundamental tactics that all good project managers know. Project Managers understand schedules can be compressed by expediting suppliers, overlapping work tasks, compromising on materials of construction, working overtime, shift work and relaxing specifications.

A step change in project execution speed cannot be achieved by applying more of the same type of tactics. Although these common tactics are important and all good project managers should be using them to speed up design and fabrication, they soon reach a point of diminishing returns. For instance, five mechanical technicians might fit comfortably working within one area of a module, and fabrication can be sped up if that number is increased to 7, but efficiency starts to decrease when the skid becomes crowded. Adding more personnel to that area will lead to further congestion, safety problems, and possibly an even longer schedule.

Increasing project execution speed is possible, but only through setting up the project at the beginning with a different strategy. The traditional design and fabrication set up breaks the job into pieces: full detailed design and costing first, followed by procurement and fabrication. Significant schedule reduction can be achieved doing these activities in parallel. The philosophy that leads to a step change in reducing schedules is to get access to pilot plant metrics, and locate the design and fabrication team at the same facility, which allows an Integrated Approach. A secondary, but notable benefit of the latter point is that safety can be improved by locating the design and fabrication team away from the final installation location.

The chart in Figure 1 illustrates the differences between a traditional engineering and fabrication approach and the Integrated Approach of a specialty fabricator. In an article written for Chemical Engineering magazine, Rich Palluzi sums up the main reason for the differences in a few words:

> "rarely is there enough information available for pilot plant projects to make an intelligent estimate of these costs, unless one has a great deal of historical data." P. 52

Palluzi, R. (2004). Pilot plant cost estimating: Make intelligent use of contingency. *Chemical Engineering, Volume (Issue),* 52 – 56.

Mr. Palluzi was discussing contingency in this particular article, however the same principle applies to working out the costs and schedule associated with detailed design, procurement, fabrication and factory testing of a pilot plant. Companies that are untested in the design and fabrication of modular pilot plants, or even modular commercial plants, do not have the metrics available to estimate costs with limited data. For that reason, a company lacking this experience and data is limited to completing full detailed design before entering a bidding phase for fabrication, as illustrated in the Traditional approach in Figure 1.

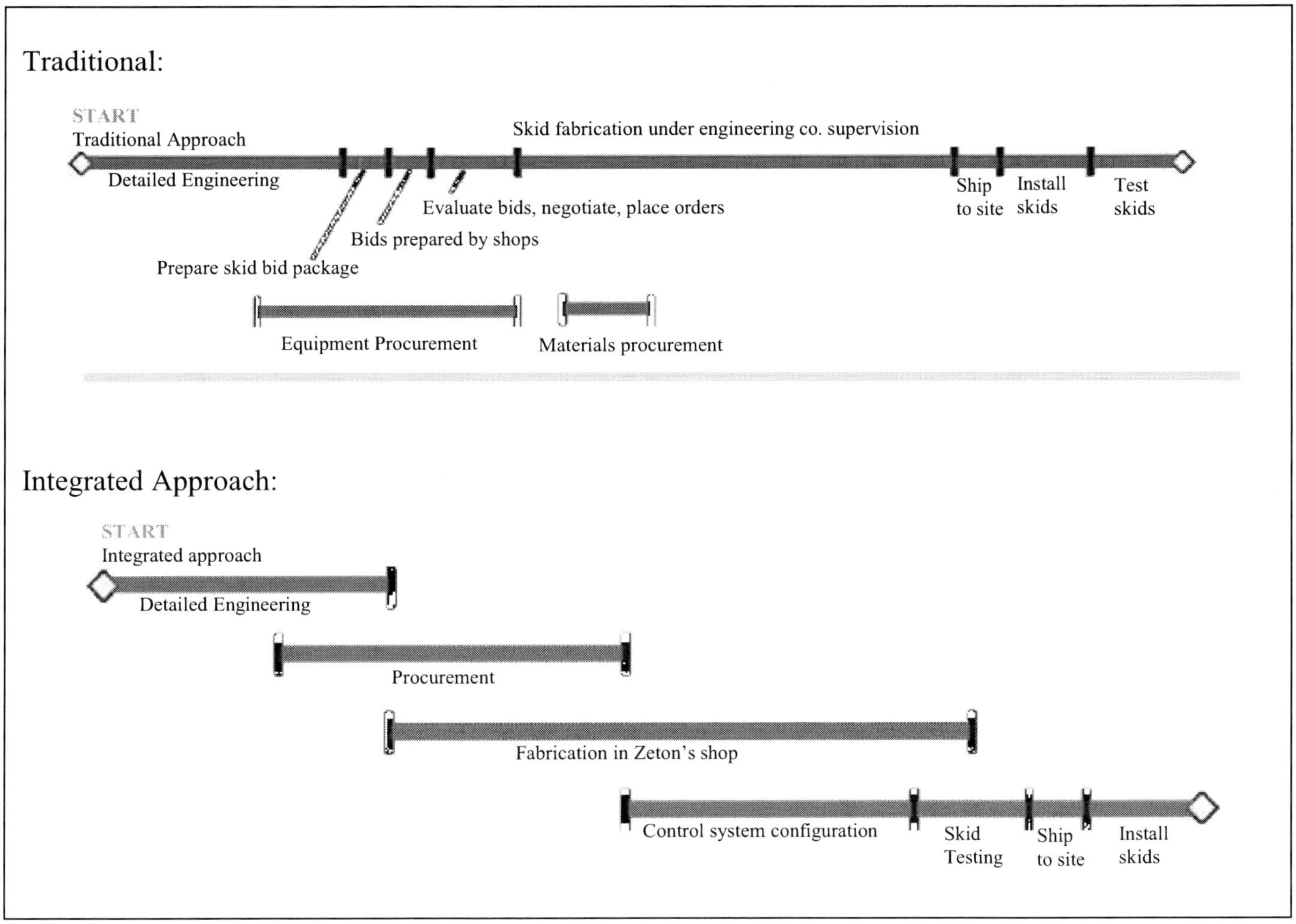

Figure 1: Traditional versus integrated design and fabrication approach.

On the other hand, specialty modular pilot plant companies have collected hundreds of data points. Once prices have been fixed in three main categories, major equipment, instrumentation, and hand valves, and some other limited information collected, the specialty company can work out an accurate price that can form the basis for a lump sum contract. Access to pilot plant metrics impact schedule directly by limiting the amount of detailed engineering that needs to be completed before the next phase.

Although all detailed engineering needs to be completed at some point throughout the project, only about one-half of it needs to be completed at this early stage. Depending on the size of the project, the first phase called Basic Engineering can last as little as 8 weeks or as long as 7 months. In other words, reducing the amount of time required by one-half cuts this part of schedule down by 8 – 28 weeks depending on the size of the project.

Of the remaining amount of detailed engineering work to be done, about one-half can be overlapped with procurement and fabrication as illustrated in Integrated approach in Figure 1. As mentioned above 8 – 28 weeks can be saved in Basic Engineering by having access to a large volume of metrics collected over hundreds of jobs. Nevertheless the time savings gained during the first stage of detailed design does not turn into an equal savings on the overall schedule. The final decrease will be about ¼ of total detailed design. So about 4 weeks on a small project and 14 weeks on a large project.

The second step change in schedule is realized by locating an experienced pilot plant design and fabrication team under the same roof. The most important part of this philosophy is that communication is greatly improved. As can be seen from Figure 1, part of detailed design is overlapped with fabrication, which is done by providing 'just-in-time-design' to the fabrication shop floor.

Close involvement of the engineer throughout fabrication ensures fabrication continues unimpeded. The point is not about cracking the whip to ensure all mechanical technicians are working hard. Rather it is for the engineer to act as a resource to the fabrication team, so obstacles can be removed from their path as the project proceeds. If any questions or uncertainty arises they are immediately answered and resolved The end result: less standing around waiting for orders and issues to be resolved; constant work flow and reduced schedule; better quality work – less miscommunication.

Another distinction between the traditional approach and the integrated approach is the potential for rework. During the traditional approach, the engineer might make a trip to the fabrication site once per week. The mechanical technicians in this case are not be specialized. This combination infrequent communication and non-specialized labor creates a high probability for incorrect installation and rework.

The mistake is multiplied further when it happens shortly after the engineers visit to the fabrication shop. If weekly visits are the norm, the next inspection of the work will happen the following week. A lot of work can take place in one week, and much of it will be dependent on the area where the mistake was made. Undoing this work to fix the error adds schedule risk. The reason traditional projects prepare a full and complete detailed design package prior to fabrication is to eliminate the need for constant engineering supervision. The drawback is that schedule suffers.

As mentioned earlier, with the integrated approach engineers who understand process concerns are on the shop floor throughout the entire project duration, and available to answer questions. This availability not only removes informational obstacles for the fabrication team, it also reduces assumptions and mistakes. Mechanical and electrical technicians are more likely to install parts the right way when they have fast and accurate answers. Errors are less likely to occur in the first place reducing schedule risk.

Despite the project manager's best efforts, mistakes happen regardless of the quality of the fabrication plan and schedule. The difference between the two systems is that when errors are made during the integrated approach, they are caught and fixed immediately. Preventing rework that would happen as a result of missed mistakes saves schedule further, and can be the difference between meeting and missing the deadline.

Planning for the deadline involves gaining access to metrics for building pilot plants. Specialty companies using an integrated approach have designed built hundreds of pilot plants. Through this experience, large databases can be developed that collect and track the mechanical and electrical technician hours required to complete tasks for various types of pilot plants. The final hours predicted to complete fabrication are based on a crew that specializes in building pilot plants. In many cases, the technicians have worked on as many as 100 pilot plants. When contrasted with a traditional fabricator that does not have experience with modular pilot plants, it is reasonable to expect more hours to be required from their fabrication team. Quoted schedules from specialty pilot plant companies can be much tighter based the specific skills of the fabrication team.

Further to speeding up fabrication, the integrated approach also makes it possible for the technology developer to complete other important activities in parallel. Preparing the site to receive the pilot plant modules prior to their arrival is very realistic. Also, while the plant is being built, research and development efforts continue in parallel, often leading to changes in the process technology that must be incorporated into the plant. This requires a flexible, disciplined and integrated approach between the specialty pilot plant company and the technology developer, such that the project continues moving towards completion.

The last step that impacts schedule is testing the entire integrated pilot plant. The most important date for schedule driven projects is the date the plant starts producing product. Clients sometimes request the plant be installed at site prior to testing, but this is not recommended. It might make sense from the perspective of a funding cut off date, or some other public relations announcement, but it rarely speeds up commissioning. In some cases, it is difficult to complete a full test. For instance if 100 ft towers are built in two pieces, they cannot be stacked and tested in the same orientation as the installation location. In other cases, the entire plant can be connected and run in the proper orientation. In either case, a factory acceptance test with simulating fluids should always be completed in the shop to the maximum extent possible to reduce the overall time to start up.

At Zeton, the engineers that design the units are always involved with the tests, and they are mainly looking for proper operation of rotating equipment and instrumentation. When a part of the plant is found to be operating below the specified limits, it can be fixed faster in the shop than it can be on site. The shop is equipped with the right tools such as overhead cranes, and a clean, well-light, work space. Testing can be completed quicker, because technicians are experienced from their work with many other pilot plants. Vendors local to the fabrication facility can fix their products immediately. In other cases where the vendor is not local, malfunctioning equipment can be removed and returned to the vendor while the plant is being trucked to site. There is a good chance this equipment can be replaced and installed prior to commissioning. Although the plant arrives later at site, testing at the fabrication facility always reduces the overall schedule by saving time during commissioning.

When all of this work is completed in the shop of a specialty pilot plant company, overall project safety is improved. Roland Meissner III commented in his article 'Modular versus conventional construction: Choosing the right path'

> "offsite prefabrication avoids plant hazards such as the performance of hot work near adjacent operational units, and can overcome access restriction that arise in an operating process plant" p. 49.

Meissner III, R. E. (2003). Modular versus conventional construction: Choosing the right path. *Chemical Engineering, September 2003, Volume, (Issue)*, 46 – 52.

Also, building the plant off-site in a controlled indoor environment improves safety, especially on larger projects. Projects stick-built outdoors must contend with rain and snow, and these slip hazards are avoided in a controlled environment. Overhead cranes in the fabricators shop provide an element of control that improves safety over stick built projects.

The answer to the question "how can I get my pilot plant faster" is to find a way to gain access to pilot plant metrics, and ensure a specialized design and fabrication team is located at the same facility, allowing design and fabrication to proceed in parallel. Without a large database of pilot plant design and construction metrics, detailed design needs to be completed in series with fabrication. With the metrics, 4 to 14 weeks can be cut out of the schedule as a result of overlapping about ¼ of the detailed design tasks with fabrication. The second step of locating the team under the same roof is a prerequisite to overlapping detailed design with fabrication. Just-in-time-design will not work any other way, because communication needs to be daily and in person. Reduced errors and less rework are inherent advantages of this structure, which have a sizeable impact on improving schedule, or at least hitting the target completion date. When design and fabrication is outsourced to a specialty pilot plant company, research and site infrastructure work can proceed in parallel, with an added benefit that safety can be improved.

ExxonMobil
Research and Engineering

Best Practices in Heat Tracing Pilot Plants

Richard Palluzi
ExxonMobil Research & Engineering Company
Clinton, NJ
April 4, 2012

Introduction

- Pilot plants are often required to operate at elevated temperatures
- Keeping all piping and components heated can be challenging
- Some industrial methods are too difficult or costly on the pilot plant scale
  - Wrapping is rarely possible in smaller size tubing
  - Many traces are larger than the tubing leading to poor contact area
  - Piping complexity leads to proportionally more heat sinks
    - Supports
    - Valves
    - Branches

Major Types of Heat Tracing

- Electric
  - Slightly more expensive than other methods
  - Easier to install, maintain, modify, and reuse
  - Types:
    - Self Limiting
    - Constant Wattage
    - Tubular Resistance Heaters
- Steam
- Hot oil
- Heated enclosures

BEST PRACTICES

Self Limiting

- Make sure that the heat output is at the desired operating temperature
  - Tracing output is a function of operating temperature
  - Adding tracing may not help
  - Recognize that the maximum temperature they can reach is usually significantly less than the published values
- Start up requires significantly more power
  - Restricts lengths per breaker often requiring multiple zones for larger applications
  - Starting at a lower temperature may not be possible!

Self Limiting

- Experience has shown that significant conservatism is usually required
  - Poor heat transfer from tracing to tubing due to geometry
  - Plan on 20-30% extra power
- Wrap generously
  - Include supports, dead legs, gauges, instruments, etc.
- Heat transfer cement and heat transfer tapes rarely very effective
  - Help but rarely cost effective vs. simply using a higher wattage tape

Constant Wattage

- Quality varies widely from good to poor
  - Lower cost laboratory supply house models are often non-uniform and have a low service life
  - Heavier duty (and more expensive) commercial versions often last much longer
  - In house testing is recommended
- Connected components can get so hot as to create a burn hazard
  - Standoffs or insulation barriers are usually required
- Installation in an electrically classified area is very difficult
  - Heater wires almost always exceed autoignition temperatures no matter how low the control temperature
  - Any leaks will have to pass an ignition source

7

ExxonMobil
Research and Engineering

Potential Solutions

- Purchasing higher voltage tape and running it at lower voltage (e.g. 120 VAC on a 240 VAC tape)
  - Depends on numerous factors so testing before use is recommended
  - Be careful that additional insulation will not inadvertently lead to higher temperatures
- Limiting the voltage (variable transformer) or amperage (phase angle fired SCR)
  - Many SCR's cannot limit either and the heater wire will reach operating temperatures but for a shorter time which is not safe
  - Again testing is recommended
- Putting a thermocouple on the heater wire is desirable but almost impossible in practice

8

ExxonMobil
Research and Engineering

Constant Wattage

- Poor wrapping techniques can create hazards
  - Crossing or doubling can lead to a fire
  - Shorting to or even through piping has occurred
  - Easily confused with wrap on insulation when removed and replaced
  - Need to leave clearly defined spacing between wraps

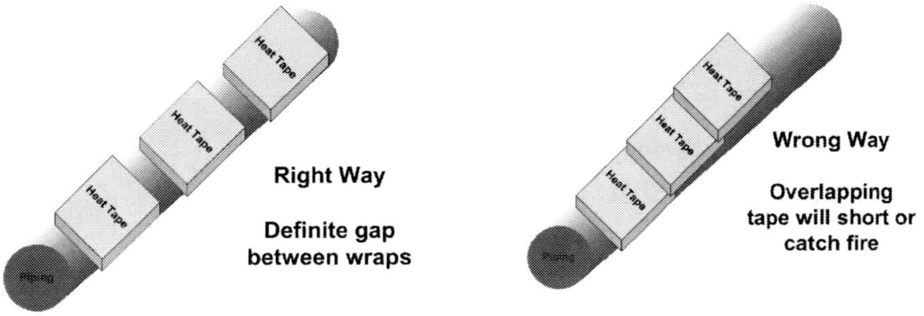

9

ExxonMobil
Research and Engineering

Constant Wattage

- Tapes are both non-uniform and often high heat capacity which results in significant temperature variation
  - Suggests that use should be limited to cases where higher temperature variations can be accepted
  - Some trial and error testing may help minimize
    - Rewrapping with more spacing between wraps
    - Adding or removing insulation in specific areas
    - Time consuming and often of limited effectiveness
- Burn hazards are very common
  - Often difficult to identify in advance
  - Usually found only after a burn is received
  - Extreme care in initial testing and operation is required
  - Thermal imaging devices can be very helpful in identifying
- Heat transfer cement and heat transfer tapes rarely very effective

10

ExxonMobil
Research and Engineering

Tubular Resistance Heaters

- Connected components can get so hot as to create a burn hazard
  - Standoffs or insulation barriers are usually required
- Laying the tubular heater against the piping is recommended
  - Wrapping can break the heater if the bend radius is too close
  - Reuse (wrap-unwrap-rewrap) usually leads to fatigue failure
- Numerous, closely space ties are recommended
  - Heaters tend to pull away when hot, leading to poor heat transfer
  - Every 2-4" is recommended
- Installation in an electrically classified area is very difficult
  - Exterior surface almost always exceeds autoignition temperatures no matter how low the control temperature
  - Any leaks will have to pass an ignition source

11

ExxonMobil
Research and Engineering

Potential Solutions

- Purchasing higher voltage tape and running it at lower voltage (e.g. 120 VAC on a 240 VAC tape)
  - Often more successful due to averaging effect of heater sheath and internal fill
  - Very dependent on actual installation so testing after installation is required
  - Be careful that additional insulation will not inadvertently lead to higher temperatures
- Limiting the voltage (variable transformer) or amperage (phase angle fired SCR)
  - Many SCR's cannot limit either and the heater wire will reach operating temperatures but for a shorter time which is not safe
  - Again testing is recommended

12

ExxonMobil
Research and Engineering

Potential Solutions

- Putting a thermocouple on each tubular heater and controlling to below the autoignition temperature
  - Requires separate control and alarm thermocouples and controllers
  - Finding the hottest point on the heater may require testing after installation

13

ExxonMobil
Research and Engineering

Tubular Resistance Heaters

- Tubular heaters are difficult to wrap around supports, instruments and other heat sinks so cold spots are more likely
  - Care necessary in piping design
    - Insulation breaks
    - Supports around insulation instead of piping
  - The first application is usually the best so advance planning and care in installation is more effective than trying to correct problems
- Burn hazards are very common
  - Often difficult to identify in advance
  - Usually found only after a burn is received
  - Extreme care in initial testing and operation is required
  - Thermal imaging devices can be very helpful in identifying

14

ExxonMobil
Research and Engineering

Tubular Resistance Heaters

- Heat transfer cement can be effective
  - But only if the heater's thermal expansion of the heater is controlled or the cement usually has a short life due to breaking off
- Heat transfer tapes usually are marginal
  - Most have difficulty due to high heat fluxes
  - Many begin to separate from heaters over time reducing effectiveness

Steam and Hot Oil

- Modifications are costly
  - Plan for extra zones
  - Deliberately undersize zone lengths so additions are possible
- Restrict zone lengths to reasonable runs and not just available pressure drop
  - Long zones often lead to cool area at the end when all the steam has condensed or the hot oil's ΔT is almost exhausted
  - Hydraulic problems can arise
- Leakage is always a potential problem
  - Keep joints outside the main insulation so they can be more easily accessible for inspection and repair
  - Weld or braze connections instead of using fittings to minimize leakage

Steam and Hot Oil

- Heat transfer cements can improve effectiveness
  - Particularly noticeable on larger components (vessels, tanks, etc.)
  - Less effective on smaller piping
- Consider the use of jacketed lines instead of tracing
  - Significantly more effective due to excellent heat transfer area
  - Fittings need to be bypassed
  - Most cost effective for longer runs with few or no fittings

Steam

- Always provide a dedicated steam trap for each zone
  - Ganging zones on a common trap always leads to some areas not working properly
- Recognize that steam traps will require some routine maintenance
  - A sudden "improvement" in a zone is often an indication the trap is bypassing live steam

Hot Oil

- Most heating fluids break down over time
  - Consult the manufacturer's recommendations for testing/replacement timing
- Many high temperature fluids are or become hazardous
  - Evaluate how to safely change out degraded fluids before construction
  - It may require additional piping and/or safeguards (ventilation, closed transfers, etc.)

Heated Enclosures

- Easy to modify process as nothing is affixed to the piping
  - Difficult → impossible to retrofit so evaluate carefully before construction
- Commercial strip heaters provide longer life and lower cost
- Provide air circulation to minimize temperature differentials
  - Simply relying on heat rising by putting the heaters on the base of the box often effective for small enclosures
  - Placing inlet holes/strips at the base and outlet hole/strips at the top often effective for moderate size and/or lower temperature enclosures
  - Air amplifiers (higher temperatures) or fan (lower temperatures) are required for larger enclosures

ExxonMobil
Research and Engineering

Heated Enclosures

- Evaluate how valves and similar components will be actuated
  - Requires additional care in layout to make sure all components are accessible from outside
  - Occasionally opening a door to turn a valve may sometimes be acceptable
  - Extending valve handles is normally the preferred solution
    - + Some additional extension for cooling may be required
    - + Be careful that the attachment mechanism cannot loosen with repeated heating/cooling
  - Remote actuation may sometimes be required
- Meeting area electrical classification is relatively straightforward

21

ExxonMobil
Research and Engineering

Potential Solutions

- Purchasing higher voltage heater and running it at lower voltage
  - Testing before use is recommended but usually more successful due to averaging effect of heater sheath and internal fill
- Limiting the voltage (variable transformer) or amperage (phase angle fired SCR)
  - Many SCR's cannot limit either and the heater will reach operating temperatures but for a shorter time which is not safe
  - Again testing is recommended
- Purging per NFPA-496
  - Allows temperatures above the autoignition temperature
  - Inert gas purging may be required
- Limiting the surface temperature of the heater

22

ExxonMobil
Research and Engineering

Surface Temperature Limitation

Controls heater power when final temperature is reached. Typically a standard PID control.

Turns heater off whenever surface temperature approaches a preset value < AIT. Typically a non-latching alarm or on-off control.

Shuts off heater if surface temperature accidently reaches AIT. Typically a latching alarm.

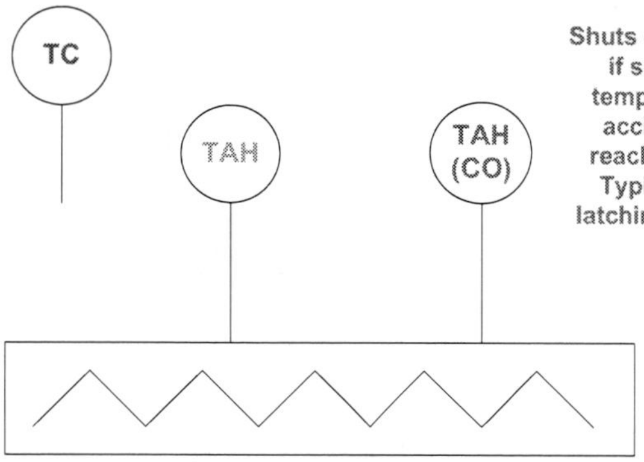

Caution

- The power delivered will be reduced proportional to the reduction in surface temperature.
- Many heaters will tend to overshoot significantly past TAH so a larger offset from TAH(CO) may be required.

23

ExxonMobil
Research and Engineering

Heated Enclosures

- Burn hazards while significantly less of a problem can still arise in unexpected places
  - Common problem areas
    - Doors and panel clearances
    - Valve handle extensions immediately next to the heated enclosure
    - Process lines immediately next to the heated enclosure
    - Structural supports
  - Usually easily found but some care in initial operation is required
  - Thermal imaging devices can be very helpful in identifying

24

ExxonMobil
Research and Engineering

Less Effective Approaches

- Preheaters
  - Usually only viable if the heated section is very short like a reactor
- Heat lamps
  - Limited to low temperatures and general purpose areas
- Forced hot air systems
  - Limited to very low temperatures or system complexity increases significantly

Safety

- Make sure the method you are considering can meet the required area electrical classification
  - Self limiting and steam are generally inherently safe
  - Constant wattage and tubular resistance tracing require careful over temperature protection
  - Heated enclosures are recommended for higher temperatures
- Expansion and contraction following heating and cooling can lead to routine process leakage
  - Consider eliminating joints by welding or single lengths of tubing
  - Determine if keeping the pilot plant hot all the time is required
  - Make sure routine leak testing is a part of the start up procedure and is well thought out and carefully done

Safety

- Personnel protection needs consideration
  - Maintaining proper insulation requires routine attention
  - Proper wiring and grounding is critical
  - Lockout/tag out provisions require enforcement
  - All electric tracing requires GFI protection by the NEC

ExxonMobil
Research and Engineering

Design

- Be realistic about how to provide over temperature protection
  - Almost always requires post construction testing (and modification!)
- Be skeptical of published heat loss values based on larger installations, components or piping
  - Higher density of valves and supports often lead to much higher values
  - Insulation is rarely as well done
- Consider the need for standoffs to isolate heated components from supports
- Insulation installation is as important as the heating yet is often very poorly done

ExxonMobil
Research and Engineering

Operations

- Carefully review the need for access to the process piping
  - Lock out and tag provisions
  - Removal and reattachment issues
  - Re-insulation
- Consider the need for additional temperature indication to confirm proper operation

29

ExxonMobil
Research and Engineering

Electric Tracing

30

ExxonMobil
Research and Engineering

Steam Tracing

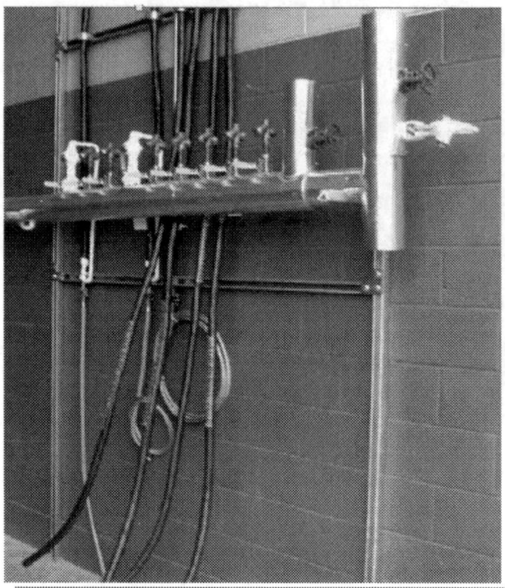

31

ExxonMobil
Research and Engineering

Heated Enclosures

32

ExxonMobil
Research and Engineering

(115c) Pilot Plant Best Practices for Safer Design and Operation

Kiran Gupte, BP Refining & Logistics Technology, Naperville, IL

BP Refining & Logistics Technology group has successfully designed, commissioned and operated numerous pilot plants in support of refining processes over the last few years. However, many of the practices adopted by the team for safe and reliable operation have been learned over more than 20 years of pilot plant experience at Naperville site. Of course, over the years the management practices have evolved with newer methodologies. In addition to the old and reliable "Process Safety Review (PSR)" process, the newer "Management of Change (MOC)" procedures have been implemented to further strengthen the safe operations. An "Inherently Safer Design" is at the heart of a new pilot plant during the conceptual design phase. The importance of "Design Phase" can not be over emphasized from the cost perspective to eliminate any cost overruns due to changes that may be needed to correct undersized or wrong equipment. This presentation will also highlight the challenges of designing and operating small scale equipment that are needed at the pilot plant scale. The pilot plant designers, operators & engineers need to appreciate the differences between pilot plant and commercial scale in order to fine tune the standard practices for safer design and operation. The attention to "Operating within the Design Envelope" while utilizing Task Hazard Risk Evaluations and Mitigations is important for safe operation with zero incidents. All these aspects are an integral part of the overarching Operations Management System. The safe practices developed are applicable to all types of equipment including small-scale catalyst screening reactors, small-medium autoclave reactors, distillation apparatus, and large pilot plant and demonstration reactors handling a broad range of hydrocarbon feedstocks.

Mini- and Micro-Scale Units in Process Development

Joseph B. Powell, Ph.D.
Shell Chief Scientist – Chemical Engineering, Houston, TX

Introduction

As bench scale and pilot units increase in size, safety risks from fire, explosion, or toxic exposure increase, as does overall footprint, costs, and required staffing. Time required for implementation increases dramatically upon progression from micro scale screening, to conventional pilot scale operation. A key question in process development is the minimum scale needed to reliably obtain information needed to assess process viability. A second critical assessment is to determine what scales and steps for pilot and demonstration facilities would be needed, if the process is found to be viable, in order to de-risk a commercial investment?

Quite often, the optimal answer to these questions entails one or more readily reconfigurable mini- or micro-scale units to allow rapid assessment of viability of key process steps, and some degree of integration between steps. Often, a further advantage of the micro- or mini-scale units is the ability to build and run multiple tests in parallel, which speeds the rate of determining the optimal process concept. Optimally, a larger scale pilot can then be designed and constructed to provide the necessary information for scale up to commercial, avoiding an intermediate demonstration phase. By using the micro- and mini-scale units to optimize among process options, more expensive and time consuming process changes at the larger pilot scale can be minimized.

Mini- and micro-scale experiments [1-5] often suffer however from less precision, and integration between units often must be done via simulation, with calculated results used to prepare mock fluids of anticipated steady state compositions as feeds. Reduced precision is typically best accommodated via use of side-by-side or consecutive comparison with known systems as a benchmark. Contacting history and time constants for all rate-dependent mechanisms will in general not reproduce commercial operation, but underlying phenomenon can be isolated, magnified, and studied, with appropriate experimental design [5]. Significant time is required in design of experiments, in order to draw out, understand, and magnify the desired parameter effects, while also eliminating possible anomalies from operation at the micro- or mini-scale.

Laboratory Catalytic Reactors

Catalytic fixed-bed reactors have historically been sized via minimum tube diameter to prevent channeling [6], which for commercially-relevant catalysts typically means one-inch diameter or larger tubes packed with 10 – 100 grams of catalyst. Footprint and required throughputs for these bench scale or mini-pilot units is relatively large. In recent years, use of catalyst particles immersed in finely divided diluent [7] has been established as a viable method for screening these systems, especially for hydrocarbon solvents, and especially where side-by-side or sequential comparison testing is conducted.

Bubble columns or stirred reactors are conveniently studied via laboratory stirred reactors, where catalyst baskets can be used to prevent fixed-bed catalyst attrition. Scale-up from stirred laboratory reactor to commercial bubble column can be done from first principles, without having to re-pilot bubble column performance. Similar scale-up from stirred reactor to commercial loop reactor can be anticipated as feasible, if potential for erosion in presence of suspended solids is anticipated.

Separations

Relatively less attention has been given to scale-down of separations [1]. Often, a crude separation step is needed during process development to provide *integration* between process steps including multi-step reaction processes. Adsorption, ion exchange, or even crystallization can be conducted at the micro scale via use of tumbling or stirred vials [1]. Simulation of thermal separations (distillation) at the micro scale is more difficult, unless a simple single-stage flash is involved. "Mini" scale batch distillations can be conducted to determine separation factors relative to known components, to calibrate interaction parameters needed for numerical simulation. One alternate approach is to not do the separation, but to use analytical characterizations and kinetic modeling to examine the integration between first and second step reactions.

Recent success was obtained in implementing micro-scale separation via use of sintered metal filtration to assess rates of dissolution, and eliminate plugging of fixed-bed catalyst, for a micro-scale coupled pilot unit. Periodic manual blowback to remove entrained solids provided both analytical information on dissolution kinetics from step-1, and provided viability for directly coupled fixed-bed reaction in step-2. Use of oversized filters allowed blowback frequency to be set at 1.5 hours vs. anticipated 15 minutes for commercial operation, in order to avoid time delay required for implementation of automated back flush, for a system involving liquid flows at 0.02 to 0.20 ml/min.

Virtual Integration Example

Figure 1 presents an example result from a study where micro-scale batch reactors were used to assess the impact of liquid phase residence times on process yields for a two-step continuous flow process, where a separate fixed-bed catalyst would be employed for the second conversion step. For the commercial process, steps #1 and #2 would be coupled in series with no intermediate cooling. A batch experimental sequence with crushed catalyst (slurry form, less than 100 microns) was employed, to test the impact of separate step-1 residence time and test for possible anomaly due to cool down required for step-2 catalyst addition. Results quickly showed a negative impact on yields from step-1 residence time, with a minor impact of the laboratory anomaly associated with cool down interval between steps. The result directed efforts to development of alternate integrated process concepts that did not require sequential two-step contacting, for maximization of yields. Reactor type and contacting history did not match proposed commercial units, but by careful design of experiments, process options could be screened in much less time than would have been required for assembly of a two-step, continuous flow integrated pilot.

Importance of Side-by-Side Comparison

A second micro-scale pilot example entailed ion exchange pretreatment, reaction, separation by crystallization, and a final thermal finishing step. Anomalous behavior in the corresponding commercial unit was diagnosed by doping impurities with feed to the initial pretreatment step. None of the separation factors or kinetics and reaction profiles matched the continuous commercial process, and only the thermal heating (and not separation) was implemented for the final thermal separation step, with analytical assessment of result. Side-by-side comparison across a sequence of batch reaction and separation steps with analytical characterization of the resulting product yields, where each step employed less than 10 grams of material, was successful in rapidly diagnosing the source of trace chemistry responsible for off-spec commercial product. Final isolated product was never produced in the lab sequence.

Over-instrumentation trap

There is a natural tendency to over engineer new experimental equipment, especially at the design phase. In one recent example, two new batch multireactor units were purchased for separate process development programs. Catalyst and reactants were manually charged to Unit A, which was also manually sampled for liquid phase GC analysis. The latter did involve automation in the form of a standard autosampler ("robot"). Unit B cost more than 5X more, and was set up for automated charge of reactants, and automated liquid phase sampling with GC analysis.

More than 1500 experiments were completed in Unit A, comprising the entire extent of study needed for project completion, before the first automated experiment was completed in Unit B. A third study added a catalytic pulse microreactor as the insert of on a GC equipped with autosampler, such that reactants could be varied by simply manually changing the composition of feeds to autosampler vials. This unit produced results within one month of concept.

The lesson from these examples is that new automation can come at a significant cost in time and instrumentation. If the experiments will be long term repetitive across multiple projects, then automation can be warranted. Process development often involves streaming of different types of experiments and equipment for different projects, and hence the leverage potentially achieved from automation is often offset by changes in process and apparatus configuration.

Summary

Micro- and mini-scale experiments can often be coupled to simulate the rudimentary aspects of integrated processes, where thermal stresses and reaction times can be studied, even in the absence of precise separation or residence time history. Careful use of side-by-side experimentation and comparison with" known" behaviors validated in larger scale pilot or commercial facilities, can be used to diagnose impacts or effects of changes in process variables.

Mini- and micro-scale studies can thus be used in tandem with larger scale pilot operations, as a means for parallel study of process variables and impacts, using the larger pilot as calibration. Impacts of individual phenomenon can often be magnified to clearly measurable levels (e.g. impacts of potential catalyst poisons), which are not evident due to noise in pilot plant studies locked in to operation at design conditions.

References

1. J. B. Powell , *Shortcut Methods for Scale-up of Separation Processes*, AIChE Spring National Meeting, Tampa, FL, April 26-30, 2009
2. J. B. Powell ,*Scale-down Considerations for Pilot Plants and Bench-Scale Units,* **AIChE Spring National Meeting**, *New Orleans, LA, April 6 – 11, 2008.*
3. J. B. Powell , *Reactor Scale-down for Pilot Plant, Bench Scale, and Multi-Throughput Units*, **AIChE 2006 Fall National Meeting**, November 12 – 17, San Francisco, CA.
4. J. B. Powell, *Use of Shortcut Methods in Process Development*, **AIChE Fall Annual Meeting**, Cincinnati, OH, Oct. 30 – Nov. 4 (2005
5. J. B. Powell , *PROCESS DEVELOPMENT: IS SCALE-UP REALLY ABOUT SCALE?, 8th World Congress of Chemical Engineering, Montreal, Canada, August 23rd-27th, 2009.*
6. D. E. Mears, "Transport Effects in Laboratory Catalytic Reactors", *Ind. Eng. Chem. Proc. Des. Dev., 10*, 541 (1971).
7. S. K. Bej, R. P. Dabral, P. C. Gupta, K. K. Mittal, G. S. Sen,V. K. Kapoor, and A. K. Dalai, *Studies on the Performance of a Microscale Trickle Bed Reactor Using Different Sizes of Diluent*, **Energy & Fuels, 14 (2000)**, 701-705
8. P. M. Wilkinson, A. P. Spek, and L. L. van Dierendonck, "Design Parameters for Estimation of Scale-Up of High Pressure Bubble Columns," *AIChE J., 38*, 544-554 (1992).

Figure 1: Parallel Batch Slurry Catalyst Experiments for study of Sequential Fixed-Bed Process Concept. [Test impact of separate step-1 residence time, parametric in laboratory-required cooling interval between steps].

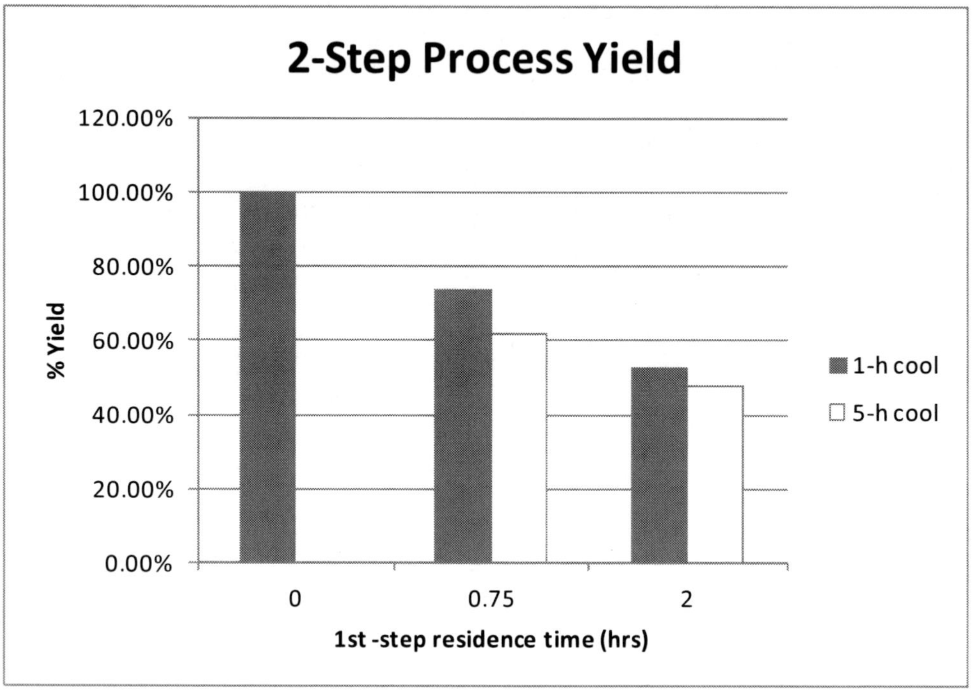

Data Integration in Petrochemical Pilot Plants by Software Optimization

Hsiang Yee Lai, Chemistry and Chemical Engineering Division, Southwest Research Institute
Theodore Wilmes, Automation and Data Systems Division, Southwest Research Institute

Introduction

Petrochemical pilot plants differ from commercial plants in that they are not continuously in steady-state operation. Petrochemical pilot plants consistently experience startups and shutdowns to allow for process design modifications. Process conditions in pilot plants vary depending on specific experimental parameters. A large amount of analytical and pilot plant process data is accumulated over time. Therefore, software development to efficiently integrate analytical and pilot plant process data becomes critical to petrochemical pilot plant operations.

The core of process engineering in pilot plants relies on the accuracy and consistency of process and analytical data acquisition. Chemical engineers incorporate this data to build multivariate statistical and thermodynamic models, which may predict methods to optimize yields and product selectivity in a pilot plant. Additionally, this data interpretation may result in system design modifications that may further improve the product characterization and overall efficiency of the system. The integration of process and analytical data into a centralized location is essential to prepare for data interpretation.

With the rise of computing technology and the increased use of process control management systems, pilot plant process data acquisition has become increasingly automated. Process data can be acquired more reliably and at a higher frequency. Several commercially available process control management software packages store historical pilot plant process data in a database format.

Analytical data, however, is not acquired through a management system. Process samples are taken by pilot plant operators and sent to chemical laboratories for analysis. A data engineer will provide a list of requested analyses for the samples. Chemists and chemical technicians analyze the samples and report the results to the data engineer. Analytical data may be received from different sources in varying formats (e.g. spreadsheets, emails, and word documents), thus increasing the possibility of communication and transcription errors. If a Laboratory Information Management System (LIMS) is in place, analytical data flow between chemical analysts and data engineers is streamlined. Data will arrive in a standard format from a single source, and is typically not stored in the same location as the pilot plant process data.

The disparity between the process and analytical data formats raises obstacles for the preparation and centralization of combined data. Data engineers end up spending more time compiling data compared to analyzing and interpreting data. Consequently, design modifications and yield optimization resulting from data analysis requires a longer turnaround time due to more time spent centralizing the data. Data flow becomes congested between pilot plant operations, chemical analysts, and data engineers. Figure 1below highlights the flow of data between each entity for a currently operational petrochemical pilot plant.

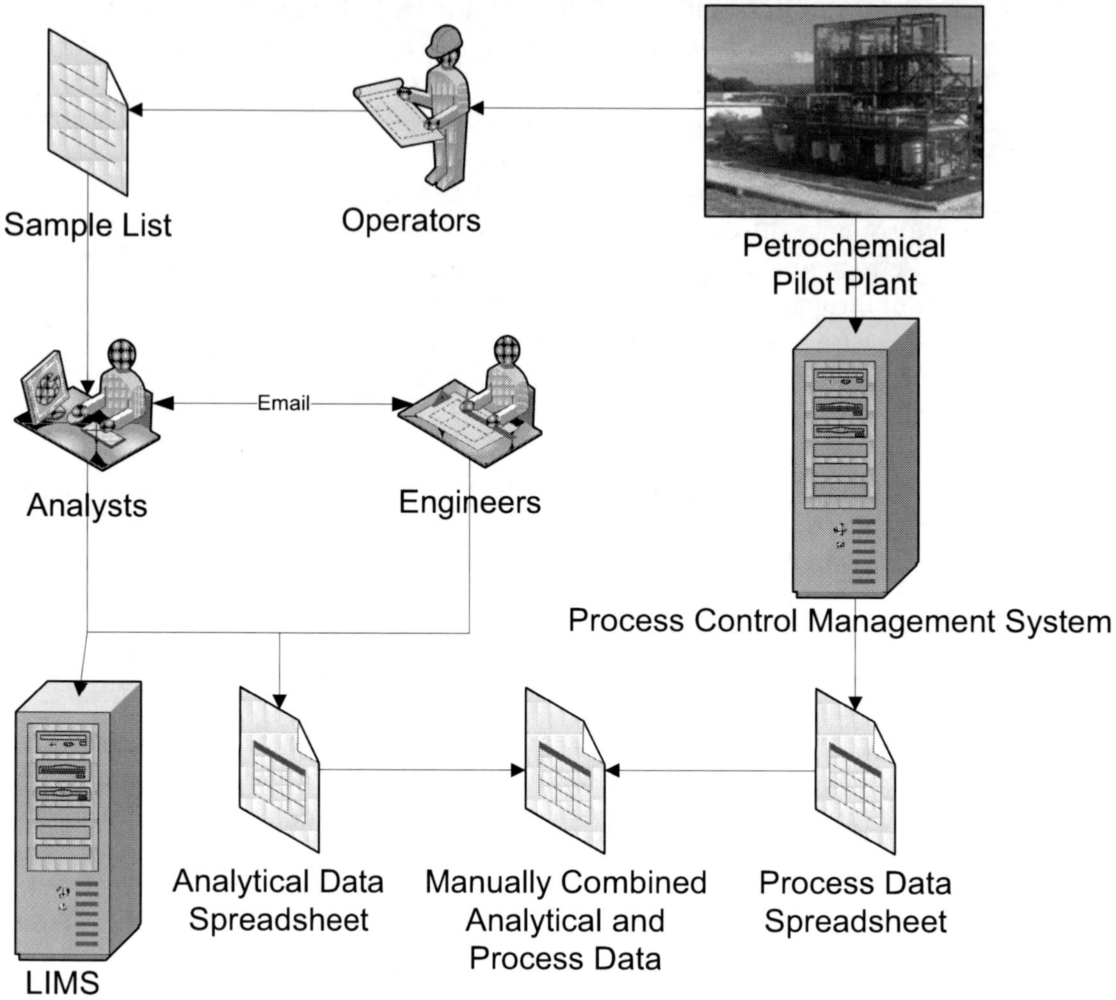

Figure 1: Original Pilot Plant Data Flow

A suggested solution to this dilemma is business process optimization. The purpose of business process optimization is to automate tasks with software where possible allowing personnel to focus on their specialties. In order to centralize and streamline the data flow between the pilot plant, chemical analysts, and data analysts, a data warehouse and customized software system were integrated into an operational petrochemical pilot plant.

Three steps were taken to achieve this: streamlining sample documentation, analytical data access, process data access.

Streamlining Sample Documentation

The existing sample documentation process was identified as an area where changes could be made to improve efficiency. Data engineer and analytical staff time was consumed by duplicative spreadsheet management and data entry. As a result, multiple points of failure were introduced where data could be entered incorrectly or simply lost, thus impacting data integrity.

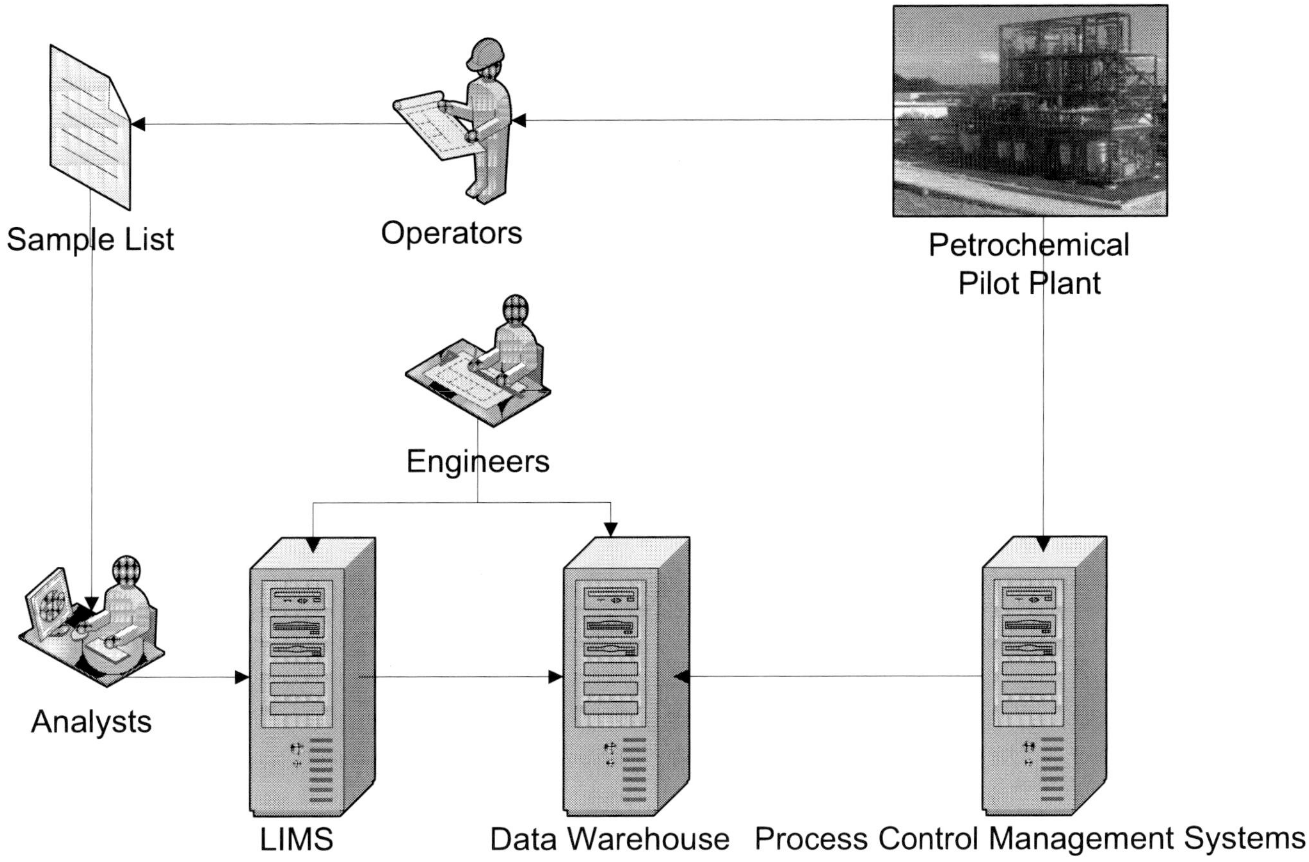

Figure 2: Data Warehouse Centered Process

The first step of the sample documentation process involved the collection of samples by pilot plant operators. Engineers would then identify and document these samples in a spreadsheet. Analytical requests would also be saved to the same spreadsheet. The samples would then be delivered to the chemical laboratories. Chemical analysts would enter the sample information into the available LIMS. The purpose was solely for sample tracking purposes. Once tests were completed, analysts would enter the test results directly into a set of shared spreadsheets. There was minimal auditing capability to track data entry, which allowed for transcription errors and a lack of accountability for said errors.

At the outset of the data integration effort, project personnel found the existing LIMS could be used to document test assignments and results. Subsequently, the LIMS was used as the analytical data source for the data warehouse.

The new process, illustrated in Figure 2, moved sample and test result entry into the LIMS. Initially, pilot plant operators collect samples, which are then picked up and documented into the LIMS by the analysts. Engineers then create task orders in the LIMS, requesting specific testing analysis for samples. Upon completion of a test analysis, the chemical analyst enters the data directly into the LIMS. The data entered into the LIMS is stored in a database on a network. Software development is required to integrate the LIMS database into a data warehouse.

The data warehouse serves as a centralized repository for the analytical and process data. Software was developed to extract analytical data from the LIMS directly into the data warehouse.

Because the LIMS data is mostly unstructured, it is first stored in a temporary holding area. An automated process parses the sample ID into discrete parts such as run, sample location, and sample number. Since sample IDs are not machine generated, the algorithm does not always recognize the sample format. As a result, sample documentation anomalies are identified. This normalization allows for query flexibility further down the road.

If the system cannot parse the sample ID, the user must remediate the issue. A custom graphical client is used for this task. The user will create the correct sample ID and then link the data of the unrecognized sample to the corrected sample identity. From this point forward, test results for this sample will be automatically recognized by the system and imported without user intervention. This step is necessary only because the LIMS does not have support for discrete sample IDs that matched the pilot plant project requirements for tracking.

Analytical Data Access

As data flows into the data warehouse, it becomes available for the engineers and other interested personnel. A custom reporting tool has been developed that simplifies report development. One of the biggest benefits of the data warehouse over just a pure LIMS solution is that the data warehouse can hold pilot plant specific information in a manner that allows for dynamic retrieval and filtering of the test data. For example, pieces of information that could only be stored as free text in the LIMS are stored in a structured format in the data warehouse. This allows system users to retrieve samples and test results based upon criteria that make the most sense to their daily work with the pilot plant. Each sample is associated with a run, location in the pilot plant, matter state of the sample (solid, liquid, gas), a unique id number, and date and time the sample was taken. The reporting tool provides the capability to filter on all of these attributes.

The custom reporting tool is not meant to replace spreadsheets or other off the shelf reporting tools, but to act as an intermediary between the data warehouse and the user. It quickly and efficiently retrieves analytical data from the data warehouse in a user-specified report format. Users first select the analytical tests that they are interested in retrieving results for and then select the samples of interest based upon specified filtering criteria. From this, the user can generate a report. The report is displayed in a tabular format similar to a spreadsheet. Data can be studied directly in the tool, or the table can be stored in spreadsheet format. In many cases, reports for outside sources still operate on spreadsheet files so this provides the necessary backwards compatibility.

Process Data Access

The industrial process control management system serves as the process data source for the data warehouse. It is important to store process data in the data warehouse to easily facilitate analyses that require an insight into analytical and process data to improve pilot plant operations. Traditionally, process data was extracted from the database in a spreadsheet format. The highest granularity of data extracted with spreadsheets was at a frequency of one minute. This would require two hours of a data engineer's attention to achieve on a weekly basis. The modified system automatically retrieves larger, bulk process data extracts, and loads them into the data warehouse. Data is extracted from the process control management system at the highest level of granularity available, which for some process variables is one read per second. As part of the import process, the process data is also stored at a one minute average granularity. This allows for greatly improved retrieval speeds and for many cases, is more than adequate resolution.

The process data is accessed primarily through a custom developed charting tool. As with the analytical client, meta-data is present in the database that allows the process data to be queried in a manner that makes sense for the pilot plant. For example, users can retrieve process data by pilot plant run as opposed to having to know the specific date range the run occurred. In addition to plotting process data, the process data client allows users to overlay analytical test results on the same charts. Calculated or derived process variables can also be computed and charted. For example, engineers can enter an equation into the system that makes use of concrete process variables and then chart the result.

Advantages and Disadvantages of Software Development for Data Integration

The overall advantages to this system are the centralization of the analytical and pilot plant process data into a data warehouse and the re-allocation of time spent compiling and preparing data for further analysis.

Previously, engineers received analytical testing results from chemical analysts in multiple formats (e.g. cumbersome shared spreadsheets, emails, pdfs). Engineers were expected to accurately compile analytical data and produce data reports to the client. Within this system, the same analytical data would be transcribed multiple times, introducing many possibilities for human error without a systematic method to audit errors.

Additionally, pilot plant process data had to be transferred from a process control management system database to large spreadsheet files by an engineer. Subsequently, these files needed to be loaded to the network and modified for further analyses purposes.

With the advent of the newly developed system, analytical data entry errors became traceable and easily remediated. Time spent organizing and preparing analytical and pilot plant process data has been reduced. Engineers are now able to concentrate on engineer-centric tasks such as analytical data interpretation and multivariate analyses of process variables. As a result, pilot plant improvements and final petroleum product development have moved at a faster pace.

The pitfall of initiating software development to optimize process engineering is to not clearly define the objective and purpose of the software. By avoiding this pitfall, the software development is more likely to be productive. Initially, the development of the data warehouse and software tools requires a large time and cost investment. This allows the engineer, analysts, and operations personnel to effectively improve the pilot plant system to produce the desired products and yields.

Future Development

Future development of the pilot plant data integration solution will extend the analytical capabilities from the first level of data retrieval to greater automation of data analysis using data mining techniques. The centralization of process and analytical data in the data warehouse simplifies the task of finding correlations between process conditions and process output quality. The primary interest is to predict process output based upon conceptual process conditions or hypothetical feedstocks. Other applications of data mining include root case analysis. Time series process data can be compared run to run to determine how process conditions may have led to degraded plant output.

(117a) Study on the Integrated Chilling Train and Mixed Refrigerant System for An Ethylene Plant

Karthik Krishnadevarajan, Jian Zhang, Qiang Xu and Ku Yen Li, Dan F. Smith Department of Chemical Engineering, Lamar University, Beaumont, TX

Study on the Integrated Chilling Train and Mixed Refrigerant System

for An Ethylene Plant

Karthik Krishnadevarajan, Jian Zhang, Qiang Xu and Kuyen Li

Dan F. Smith Department of Chemical Engineering

Lamar University, Beaumont, TX 77710, USA

Abstract

Refrigeration is one of the most important and crucial process in chilling train of ethylene plants. There are different ways to provide refrigeration for the process. The latest trend is to use a mixed refrigerant system (MRS) which has at least three components for the refrigeration task. An MRS provides a continuum of temperatures change with smaller temperature differences during the heat exchange process than the single or cascade refrigerant system. This leads to a lower energy loss in MRS and some other advantages than the traditional cascade refrigerant systems. On the other hand, chilling train in ethylene plants contains a series of multiple heat exchangers and flash drums to separate hydrogen and methane gas from cracked gas and to liquefy the C2 and heavier components. It handles process streams whose temperatures vary from 40 to -160°C, which requires significant amount of cooling duty. Obviously, the chilling train and mixed refrigerant systems should be integrated for study in terms of energy saving and system operational performance improvement. The present work deals with combining chilling train and mixed refrigerant system for an ethylene plant. Both process models are developed by ASPEN Plus. Based on the rigorous simulation model, the conceptual optimal design and operational optimization for improving the integrated system has been conducted.

Structure Optimization and Performance Study of Rotating Jetting Bed

Xiao-hua Li, Yu-min Li, Ying Wang, Zheng-fu Tong, Jian-bing Ji

(Zhejiang University of Technology，Hangzhou，zhejiang，China)

ABSTRACT

As a novel gas-liquid mass-transfer equipment, the rotating packed bed(RPB) using centrifugal force to greatly intensify the processes of mass transfer, has been successfully applied in absorption, stripping, distillation, extraction process and nanomaterial production of chemical industry. The rotating jetting bed(RJB) was newly developed with the core components of the rotor and coil-type liquid distributor. The rotor is made up of stationary disc and rotating disc which is fixed to a series of concentric rotating baffles with perforations. Compared to RPB, intermediate feeds can be accomplished, installing an inlet at the stationary disc. Moreover, RJB with multiple rotor coaxially installing in one casing can be employed. Thus it is possible for one RJB unit to perform a continuous distillation.

As the important component of the rotating bed, the rotor determines the performance of RJB. Experiments of total reflux distillation with five types of RJB rotor were carried out using ethanol-water systems at atmospheric pressure. Effects of *F*-factor, spray density, rotational speed and rotor structure on mass transfer performance of RJB were investigated. Then the gas pressure drop and power consumption of RJB with the rotor of three-layer composite rotational baffles have been investigated in detail using the system of air-water.

The experimental results show that:

（1）The mass transfer efficiency of the RJB with five types of rotor appears the maximum value with the increase of the *F*-factor, spray density and rotational speed. So there exists an optimal operating condition. The higher opening ratio of rotational baffle is beneficial to mass transfer of RJB. The mass transfer efficiency of the rotor of three-layer composite rotational baffles was highest in the five type of rotors, with 30~50 theoretical plates per meter.

（2）There exists the gas-phase end effect of mass-transfer at the outside part of the rotor. Coil-type liquid distributor can significantly improve the initial liquid distribution in the circumferential and axial direction, increasing the mass transfer performance of RJB.

（3）The gas pressure drop increased with the increase of *F*-factor and rotational speed. The spray density has a little effect on the pressure drop. Therefore, the pressure drop is correlated with gas Reynolds number, liquid Reynolds number and rotating Reynolds number. The calculated values agreed well with experimental values, and the average deviation was less than 10%.

（4）The power consumption of RJB increased with the increases of rotational speed and liquid flow rate, and decreased slowly with the increases of gas flow rate.

The ratio of the liquid power consumption to the shaft power consumption is 20%~70%, and the ratio increased with the increases of rotational speed and liquid flow rate.

（5）Among the five types of rotors, the rotor of three-layer composite rotational baffle with lower power consumption and lower specific pressure drops and high efficiency, was the superior rotor structure for RJB.

KEY WORDS: Rotating Jetting Bed，mass-transfer，gas pressure drop, power consumption

Email: Xiao-hua Li：lixh63@zjut.edu.cn

1 Introduction

Current chemical process intensification has become the field of chemical engineering research. Process intensification means on the premise of achieving production targets，use of new technologies or new equipment, greatly reducing device size, or reduce the number of devices, reduced byproducts, dramatically increasing productivity. High-gravity rotating bed is a new type of gas-liquid mass transfer equipment，the device uses centrifugal force of the rotating greatly strengthened the process of gas-liquid contact, enhancing mass transfer efficiency. Research by Rotating Zigzag High-gravity Bed(RZB) has made great progress，and widely used in chemical, pharmaceutical, light industry and other industries, but the equipment there is a large pressure drop, flux of small disadvantages。Rotating Jet High-Gravity Bed(RJB) is a new development of the unique structure of rotating packed bed，its core part is the rotor and inventory control-type liquid distributor, rotor-stator and fitted with a series of concentric circular composition。Compared with the conventional Rotating High-Gravity Packed Bed(RPB)，RJB under the only disc rotating, plate still on, you can easily set on the stator inlet and coaxial-line multiple rotors in one device, which can be applied to continuous distillation process.

2 Experiment

2.1 RJB equipment

Main structure of RJB composed of the spray-type rotor and a new liquid distributor, as shown in Figure 1. Rotor was formed stationary disc and rotating disc with a series of concentric circular rings in disc. Rotating disks dynamic embedded stationary disc notches at the top, form a labyrinth seal. A new type of liquid distributor rotor centre was installed.

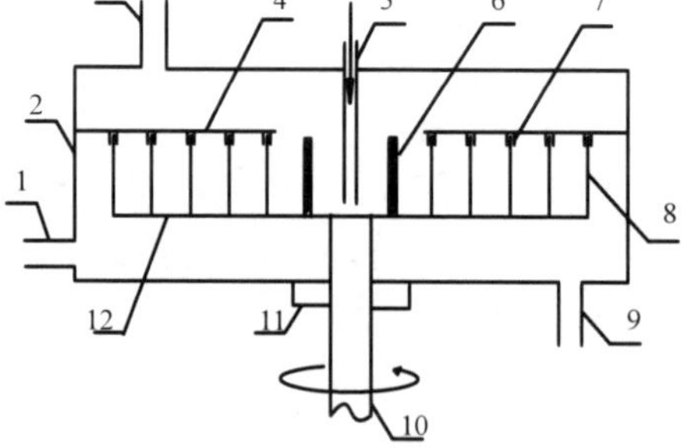

Fig.1 Scheme of RJB

1 Gas inlet; 2 Shell; 3 Gas outlet; 4 Stationary disc; 5 Liquid inlet; 6 Liquid distributor; 7 labyrinth seal; 8 Concentric rotating baffles; 9 Liquid outlet; 10 Rotating shaft; 11 Dynamic seal; 12 Rotating disc

2.2 Rotor structure of RJB

Laboratory by rotating bed shell diameter 380mm, rotor structure has five, Liquid distributor, stationary disc, etc apart from the circle of parts are the same, only five rotor circle of different, and 7 respectively with a circle of diameter 140, 160, 185, 205, 230, 250, 270mm, rotor axial height of 15.0mm. Five kinds of RJB rotor structure are:

Rotor Ⅰ: Rotating disc of pore plate with 43.5% hole rate; pore plate as shown in Figure 2 (a), as shown in the rotor I in Figure 3 (a) below.

Rotor Ⅱ: Rotating disc of pore plate with 37.3% hole rate.

Rotor III: Rotating disc of pore plate with 22.7% hole rate.

Rotor Ⅳ: Rotating disc for a pore plate with 43.5% hole rate and mesh plate with 44% hole rate, two plates close. Mesh plate for side 0.82mm, 70 ° angle diamond-shaped holes, hole rate of 44%.

Rotor Ⅴ: Rotating disc for a pore plate with 43.5% hole rate and two mesh plates with 44% hole rate, in a with a pore plate in the middle of composite plate, interval 2mm. Three types of plate with a circular hole ring structure as shown in table 1.

Table 1 Structure parameters of RJB rotor of three pore plates

Rotor	Hole diameter/mm	Distance between holes/mm	Fractional hole ratio /%	thickness of plating/mm	Material
I	1.8	2.6	43.5	0.8	stainless steel
II	1.6	2.5	37.2	0.9	stainless steel
□	1.0	2.0	22.7	1.0	stainless steel

2.3 Experimental process

As the important component of the rotating bed, the rotor determines the performance of RJB. Experiments of total reflux distillation with five types of RJB rotor were carried out in ethanol-water system at atmospheric pressure. Effects of F-factor, spray density, rotational speed and rotor structure on mass transfer performance of RJB were investigated. Then the gas pressure drop and power consumption of RJB with the rotor of three-layer composite rotational baffles have been investigated in the system of air-water. Experimental process is as shown in Figure 3.

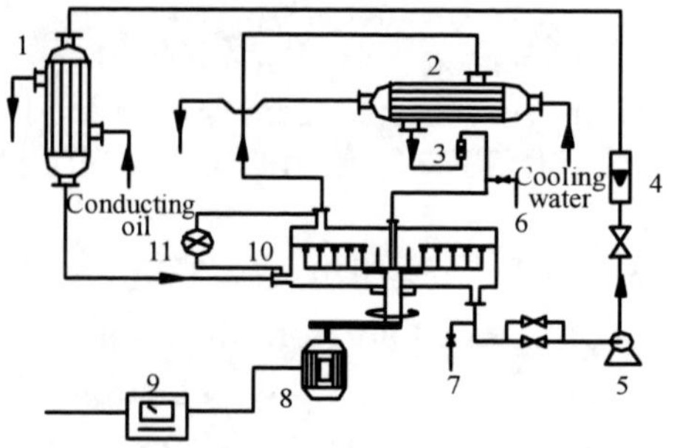

Fig.3 Experiment flow of distillation in RJB

1 Reboiler; 2 Condenser; 3,4 Rotameter; 5 Pump; 6, 7 Sampler; 8 Motor; 9 Frequency modulator; 10 RJB; 11 Pressure difference sensor**3 Results and discussion**

3.1 Effect of operation parameters on the number of theoretical plates per meter

Figure 4 shows the speed at 1000r/min and 1200r/min, different number of theoretical plates per meter of rotor with the *F*-factor, five rotor structure rotating bed along with the number of theoretical plates per meter increases the first and then slowly decrease with *F*-factor increases. Because *F*-factor increases, gas speed in rotor increases, gas liquid turbulence degree increased, conducive to pass mass strengthening; while, in total reflux condition, with *F*-factor increases, reflux liquid rate increases, packing of wetting area increases, rotor crush formed of liquid drops number increased, also conducive to pass mass of for; but *F*-factor increases, material in rotor in the of stay time variable short and rotor diameter Department gas speed increases, huge of air easy will liquid with out, caused fog foam entrainment, Both mass transfer effect of reducing device. Both compete with each other, the best operating conditions.

From figure 4 also shows that *F*-factor increase process, the number of theoretical plates per meter of rotor I is greater than those of rotor II and III, below those of rotors IV and V, mass transfer efficiency of rotor V is best. The number of theoretical plates per meter of rotor V is in 30~50.

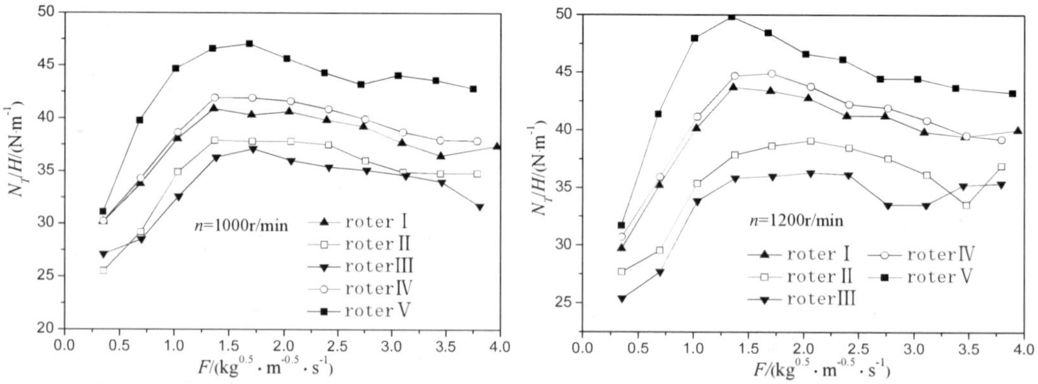

Fig. 4 Effect of F-factor on the number of theoretical plate per meter

Figure 5 shows the speed at 1000r/min and 1200r/min, the number of theoretical plate per meter is different with the spray density changes, as the spray density increases, the number of theoretical plates per meter of five kinds of rotor have maxima. Best spray density is 6~10 $m^3 \cdot m^{-2} \cdot h^{-1}$.

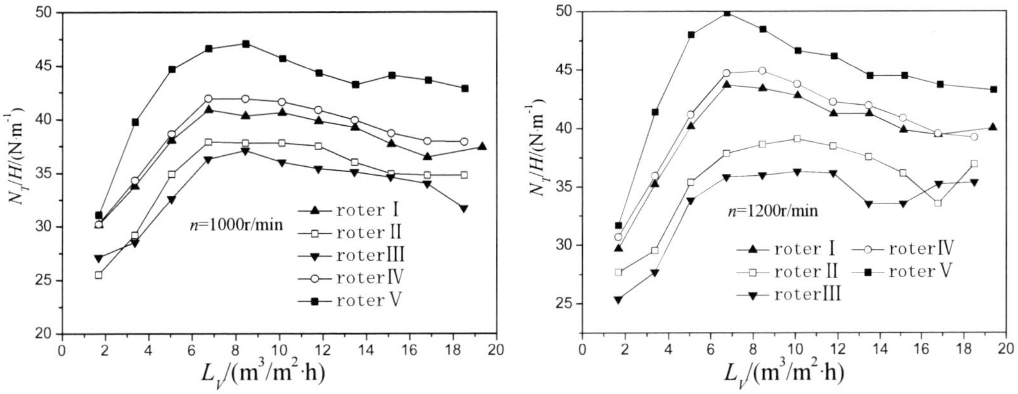

Fig.5 Effect of spray density on the number of theoretical plate per meter

Figure 6 shows at F=1.36, the number of theoretical plate per meter of rotor different with rotating speed, the number of theoretical plates per meter of rotor V increases with rotating speed increases. But at F=2.73, tends to increase slowly.

3.2 Effect of operation parameters on specific pressure drop

Figure 7 is Effect of F-factor on specific pressure drop at the time of 1000r/min

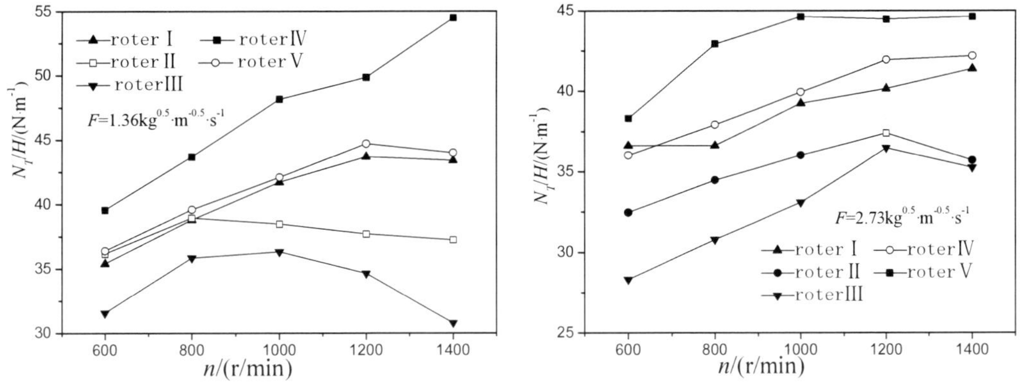

Fig.6 Effect of rotating speed on the number of theoretical plate per meter

and 1200r/min. Under operating conditions, the specific pressure drop of rotor V is reduced first and then increased with F-factor increased and those of other four kinds of rotors is gradually rising trend. When the F-factor reaches 2 quickly grow. This is because the F-factor is less than 2, F-factor and increased with the increase in pressure drop at the same time and increased with the F-factor also increases the efficiency of mass transfer, pressure drop is changing slowly. When the F-factor is greater than 2, the pressure drop along with the F-factor increases quickly grow, this is because the Jet-rotating bed pressure drop is proportional to the square of the air speed, mass transfer efficiency increase and slow decrease with the F-factor. when

F-factor is less than 2, he specific pressure drop of rotor V was lower than those of other four kinds of rotor .As F-factor is greater than 2, the specific pressure drop of rotor I significantly smaller than those of other four kinds of rotor.

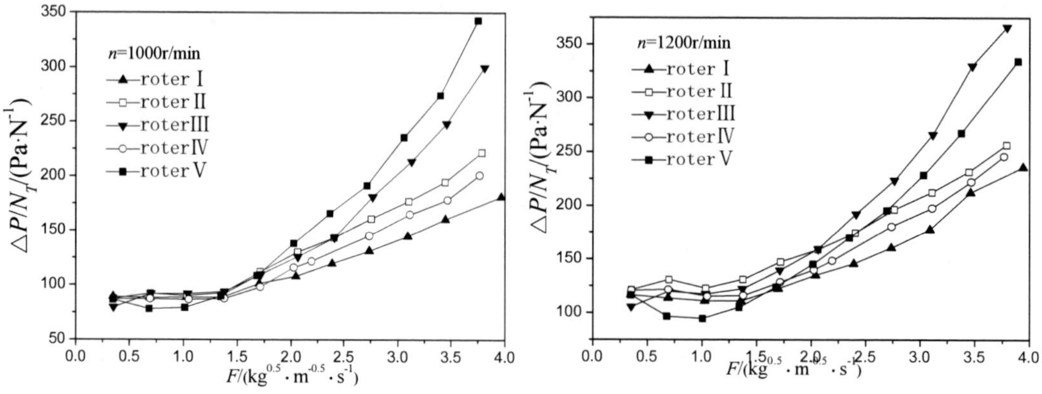

Fig.7 Effect of F-factor on specific pressure drop（$\Delta p/N_T$）

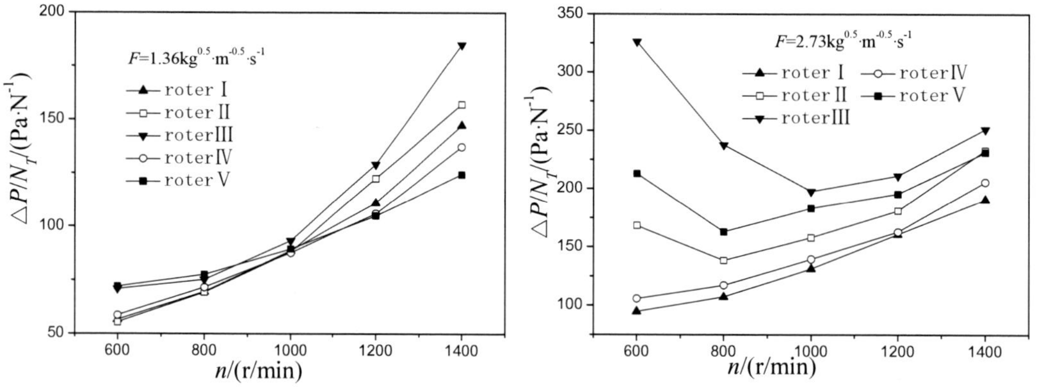

Fig.8 Effect of rotating rate on specific pressure drop（$\Delta p/N_T$）

3.4 End effect

So-called end effect mean in gas-liquid mass transfer in the process, packing layers of the rotor within a small area, due in the region in the packing of particles in gas-liquid has a very strong effect, mass effect is very good. By the predecessors of the research, in packing layers inside a end effect in liquid phase, where large volumetric mass transfer coefficient; packing these cities end effects at the outer edges of the layer area, here the volumetric mass transfer coefficient is also larger.

Using ethanol-water system, under the conditions of atmospheric total reflux, RJB distillation experiment by changing the method of injection of the inside and outside diameter of the rotor of RJB with end effects on mass transfer performance . Research results show that, spray end effects in RJB does not exist in the area, that literature reported end effects of liquid phase; outside but there is a side effect, the vapor end effects. Coil-type liquid distributor can significantly improve the liquid in weeks and weeks up in the initial distribution of the rotor, significantly improved mass transfer efficiency of RJB.

3.5 Effect of operation parameters on the power consumption of RJB

The shaft power consumption of RJB increased with the increases of rotational speed and liquid flow rate, and decreased slowly with the increases of gas flow rate. The ratio of the liquid power consumption to the shaft power consumption is 20%~70%, and the ratio increased with the increases of rotational speed and liquid flow rate.

4 Conclusion

Main conclusions are as follows:

The mass transfer efficiency of the RJB with five types of rotor appears the maximum value with the increase of the *F*-factor, spray density and rotational speed. So there exists an optimal operating condition. The higher opening ratio of rotational baffle is beneficial to mass transfer of RJB. The mass transfer efficiency of the rotor of three-layer composite rotational baffles was highest in the five type of rotors, with 30~50 theoretical plates per meter.

There exists the gas-phase end effect of mass-transfer at the outside part of the rotor. Coil-type liquid distributor can significantly improve the initial liquid distribution in the circumferential and axial direction, increasing the mass transfer performance of RJB.

The gas pressure drop increased with the increase of F-factor and rotational speed. The spray density has a little effect on the pressure drop. Therefore, the pressure drop is correlated with gas Reynolds number, liquid Reynolds number and rotating Reynolds number. The calculated values agreed well with experimental values, and the average deviation was less than 10%.

The power consumption of RJB increased with the increases of rotational speed and liquid flow rate, and decreased slowly with the increases of gas flow rate. The ratio of the liquid power consumption to the shaft power consumption is 20%~70%, and the ratio increased with the increases of rotational speed and liquid flow rate.

Among the five types of rotors, the rotor of three-layer composite rotational baffle with lower power consumption and lower specific pressure drops and high efficiency, was the superior rotor structure for RJB.

Reference

1 Ramshaw, C. (1983)., "HiGee distillation—An example of process intensification, " *Chem. Eng.*, February, pp. 13–14

2 Keyvani, M. (1989), "Operating characteristics of rotating beds, " *Chem.Eng. Prog.*, pp. 48-52.

3 Liu, H.S., Lin, C.C. and Wu, S.C. (1996), "Characteristics of a rotating packedbed," *Ind. Eng. Chem. Res.*, pp. 3590-3596.

4 Rao, D.P., Bhowal, A. and Goswami, P.S. (2004), "Process intensification inrotating packed beds(HIGEE): An appraisal," *Ind. Eng. Chem. Res.*, pp.

1150-1162.

5 Li, X.P., Liu, Y.Z. and Li, Z.Q. (2008), "Continues distillation experiment with rotating packed bed," *Chin. J. Chem. Eng.*, pp. 656-662.

6 Tong, Z.F. (2010), "Study on hydrodynamic and mass transfer performance of rotating jet high-gravity bed,"[D] Zhejiang University of Technology, Hangzhou.

7 Zheng, C., Guo, K., Feng, Y.D.and Yang, C. (2000), "Pressure drop of centripetal gas flow through rotating beds," *Ind. Eng. Chem. Res.*, pp. 829-834.

8 Guo, K., Guo, F. and Feng, Y. (2000), "Synchronous visual and RTD study on liquid flow in rotating packed-bed contactor," *Chem. Eng. Sci.*, pp.1699-1706.

9 Wang, G.Q., Xu, Z.C., Yu, Y.L.and Ji, J.B. (2008), "Performance of a rotating zigzag bed—A new HIGEE," *Chem Eng Process,* November, pp. 2131–2139.

10 Ji, J.B. , etc. (2004), "Equipment of zigzag high-gravity rotating beds," *CN Pat.*, 01134321.4.

11 Ji, J.B., Xu, Z.C., etc. (2008), "Equipment of zigzag high-gravity rotating beds", *U.S. Pat.*, 7344126.

ENHANCEMENT OF MASS TRANSFER IN CIRCULAR CONDUIT WITH COAXIALLY PLACED PERFORATED DISC AS TURBULENCE PROMOTER

D.S.Seetharama Raju[a*]**, *M.Gangadhar***[b] ***and V.Nageswara Rao***[a]

[a]*Department of Chemical Engineering, College of Engineering(A), Andhra University, Visakhapatnam-530 003, Andhra Pradesh, INDIA*

[b]*Department of Chemical Engineering, G.M.R. Institute of Technology, GMR Nagar, Rajam, Srikakulam-532127, Andhra Pradesh, INDIA*

**E-mail: sai_dandu@yahoo.co.in*

ABSTRACT

Mass transfer data were obtained in circular conduit using an electrochemical technique with a potassium ferri-ferro cyanide couple. In circular conduit coaxially placed entry region perforated disc was used as turbulence promoter. The study comprised of evaluation of mass transfer rates at the outer wall of the electrochemical cell. Mass transfer coefficients were evaluated from the measured limiting currents. The study covered a wide range of geometric parameters such as diameter of the disc (d_d), thickness of the disc (d_t) and distance of the disc from the entrance of the test section (h). The results revealed that the mass transfer coefficient increased with increase in velocity, diameter of the disc (d_d), thickness of the disc (d_t) and decreased with increase in distance of the disc from the entrance of the test section(h). Within the range of variables covered, the augmentation achieved in mass transfer coefficients were up to 2.1 fold over the tube flow in absence of promoter. The entire mass transfer data were correlated with Colburn factor 'J_D' and roughness Reynolds number. The following correlation was reported out of the study.

$J_D = 0.155(Re)^{-0.9432}(\phi_1)^{0.641}(\phi_2)^{0.4658}(\phi_3)^{-0.3775}(Sc)^{1.1521}$

Where $\phi_1 = d_d/d$, $\phi_2 = d_t/d$, $\phi_3 = h/d$ are dimensionless groups, d is diameter of test section.

Keywords: Mass transfer, perforated disc, turbulence promoter

INTRODUCTION

Researcher will always aim at the maximum output with reduced equipment size so as to minimize the unit product cost. This is particularly true in the design of electrolytic cells where the mass transfer limiting conditions exist. Use of turbulence promoter is one such technique that increases the mass transfer coefficients by several folds over the smooth flow. This disturbance continues to its effect on either side of disc and continues to a longer distance which in turn influences the wall mass transfer coefficient over the length of the tube. The study was conducted to know the effectiveness of perforated disc in augmenting mass transfer coefficient. Burns and Jachuck [1] determined liquid-solid mass transfer performance on a 30cm diameter spinning disc reactor by use of the limiting current technique. Ravi et al [2] did experiments and obtained mass transfer data on the walls of fluidized beds using an electro chemical technique with a potassium ferro-ferri cyanide couple. The disc promoter assembly was used as an internal in fluidized beds. Sherwood and Stone [3] determined the flow of a viscous fluid around disc in a pipe and the added mass of the accelerating disc and flow profiles were also presented. Padesta et al [4] obtained a dimensionless correlation for ionic mass transfer at horizontal disc electrode under longitudinal vibration. Escudie et al [5] carried out experiments in water fluidized binary mixtures of Teflon spheres, discs and rods. Venkateswarlu et al [6] studied the effect of string of discs placed co-axially in a circular conduit on mass and momentum transfer. Data on mass and momentum transfer in a circular conduit with coaxially placed perforated disc as turbulent promoter has been taken up since the data on the present system has not been reported in literature. Based on the data a semi theoretical model has been developed. Parameters covered in the present study are shown below in Table 1.

Table 1

Variable	Minimum	Maximum	Max/Min
Diameter of the disc, (d_d), m	0.025	0.045	1.8
Disc thickness, (d_t), m	0.005	0.060	12
Distance of the disc from the entrance of the test section, (h),m	0.14	0.30	2.14
Velocity, V, m/s	0.0984	0.3936	4
Reynolds number, Re	5264	21059	4
Schmidt number	891	990	1.11

EXPERIMENTAL

Schematic diagram of experimental set up is shown in figure 1. It is similar in layout to that used in earlier studies [7, 8, 9, 10]. It essentially consisted of a storage tank (TS), centrifugal pump (P), rotameter (R), entrance calming section (E_1), test section (T) and exit calming section (E_2). The storage tank is cylindrical copper vessel of 100 liter capacity with a drain pipe and a gate valve (V_1) for periodical cleaning. A copper coil (H) with perforations is provided to bubble nitrogen through the electrolyte. The tank is connected to the pump with a 0.025m diameter copper pipe on the suction line of the centrifugal pump. The suction line is also provided with a gate valve (V_2).The discharge line from the pump splits into two. One served as a bypass line and controlled by valve (V_3). The other connects the pump to the entrance calming section (E_1) through rotameter. The rotameter is connected to a valve (V_4) for adjusting the flow at the desired value. The rotameter has a range of 0 to $166x10^{-5}m^3/s$.The entrance calming section consisted of 0.05 m ID circular copper pipe with a flange and is closed at the bottom with a gland nut (G).The up-stream side of the entrance calming section is filled with capillary tubes to damp the flow fluctuations and to facilitate steady flow of the electrolyte through the test section. It is made of a graduated Perspex tube of 0.36m length with point electrodes fixed flush with the inner surface of the tube. The point electrodes are made out of a copper rod and machined to the size. They are fixed flush with the inner surface of the test section at equal spacing of 0.01m. Exit calming section is also of the same diameter copper tube of 0.5 m long, and it is provided with a flange on the upstream side for assembling the test section. It has gland nuts (G) at the top and bottom ends to hold the central tube. Two thermo wells (t_1, t_2) were provided, one at upstream side of the entrance calming section and the other at the down stream side of exit calming section for measurement of temperature of the electrolyte. Perforated disc serving as turbulence promoter is made of Nylon of various sizes with a provision to fix it rigidly within the test section. The perforated disc has four round holes of each 2mm diameter. The Perforated disc is placed concentrically in the test section. The promoters used are shown in photograph in figure 2. The details of perforated disc promoter are shown in figure3.The limiting current measuring equipment consisted of multimeter of Motwane make which has 0.01mA accuracy and vacuum tube voltmeter is used for potential measurements. The other equipments used in circuit are rheostat, key, commutator, selector switch, and a lead acid battery as the power source. The commutator facilitated the measurement of limiting currents for oxidation and reduction process under identical operating conditions by the change of polarity while the selector switch facilitated the measurements of limiting currents at any desired electrode. The circuit diagram used for the measurement of limiting currents is shown in the figure4.

Data on limiting currents for the case of reduction of ferricyanide ion is obtained for fluid flow in circular conduits in the presence of perforated disc as insert promoter. The following electrode reaction is involved.

Cathodic reduction of ferricyanide ion:

$$[Fe(CN)_6]^{-3} + e \text{ ------------- } [Fe(CN)_6]^{-4} \quad \text{...... (1)}$$

Initially blank runs are conducted with indifferent electrolyte (sodium hydroxide solution) alone to ensure that the limiting currents obtained in the subsequent runs are due to diffusion of reacting ions (Ferri cyanide ion) only. The electrolyte was pumped at a desired flow rate (through the test section) by operating the control and by-pass valves. After steady state is attained, potentials are applied across the test electrode and wall electrode in small increments of potentials (100mV) and the corresponding currents were measured for each increment. In view of the large area of the counter electrode in relation to the test electrode nearly constant potential is maintained at the test electrode. Since the potential values are not of criteria in the present study, the limiting currents were determined from the measurements of applied potentials and currents as has been done in several earlier works [7, 8, 9, 10]. The attainment of limiting current is indicated by the constancy of current with a large increase in the potential. Mass transfer coefficients are computed from the measured limiting currents by the following equation:

$$k_L = i_L/nFAC_0 \quad \text{......(2)}$$

Pressure drop measurements are also taken simultaneously using U-tube manometer with Carbon tetrachloride as manometric liquid.

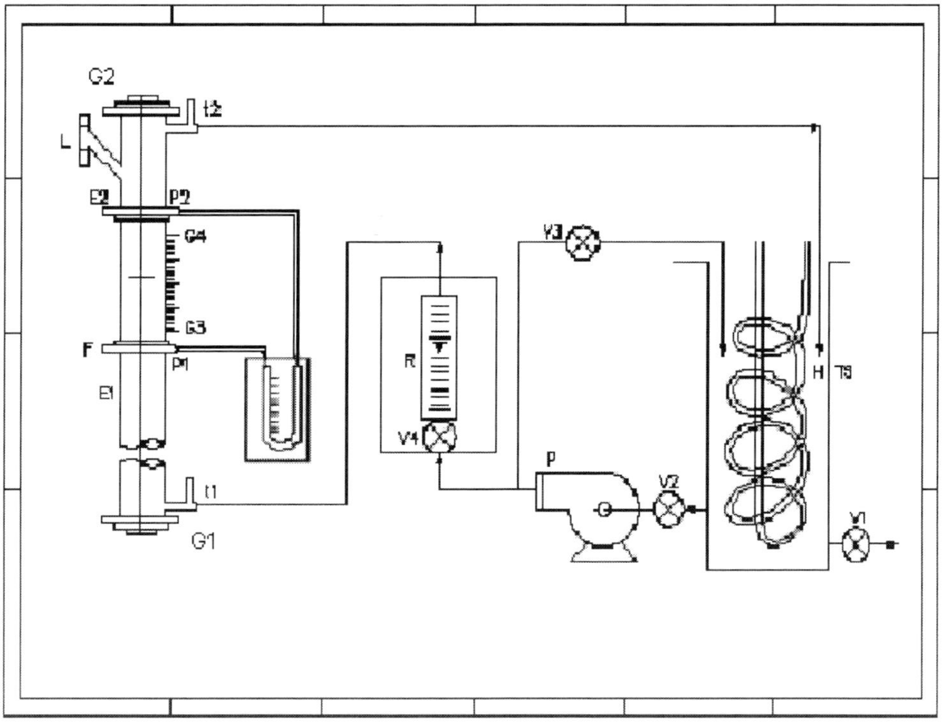

Figure 1: Schematic Diagram of Experimental Setup

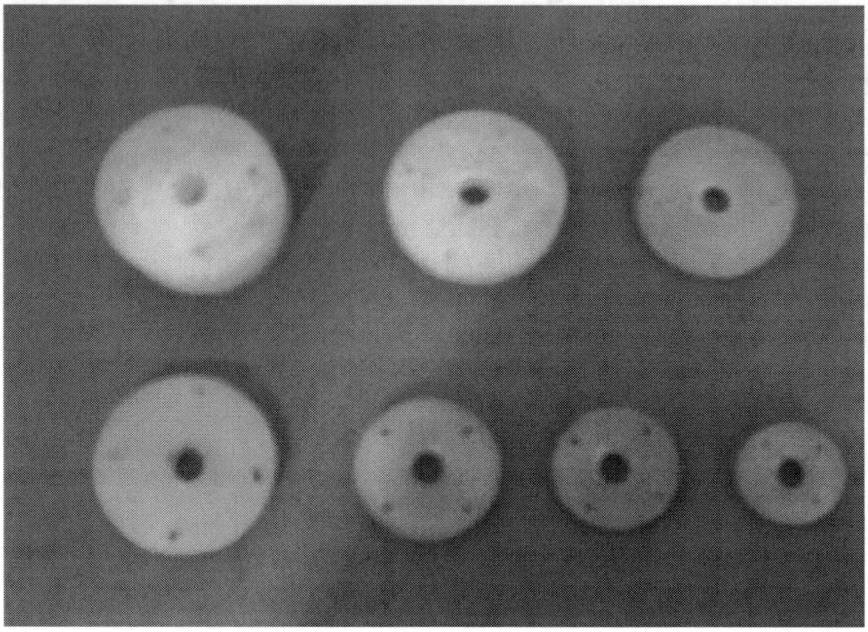

Figure 2: Turbulence Promoters

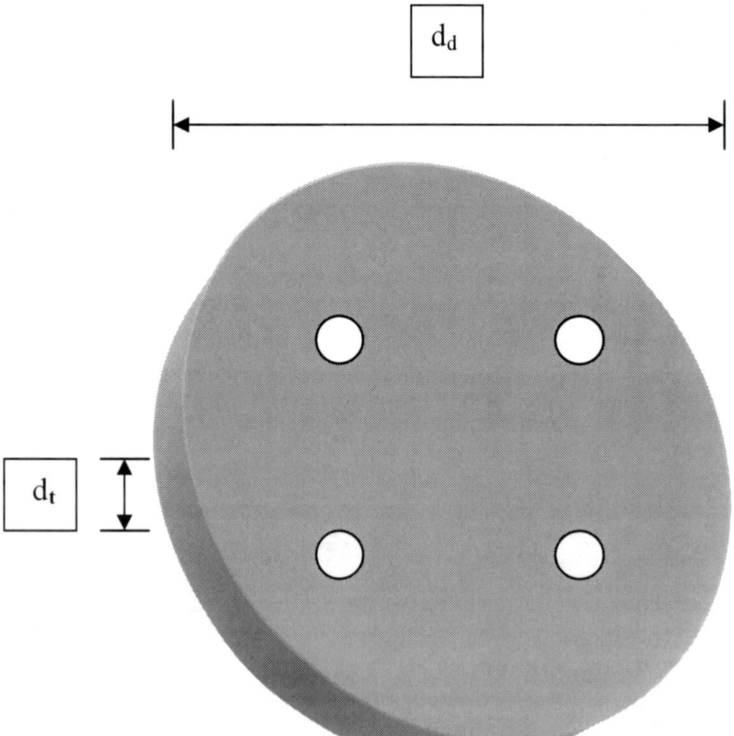

Note:
d_d = Diameter of Perforated Disc
d_t = Thickness of Perforated Disc

Figure 3: Details of Promoter

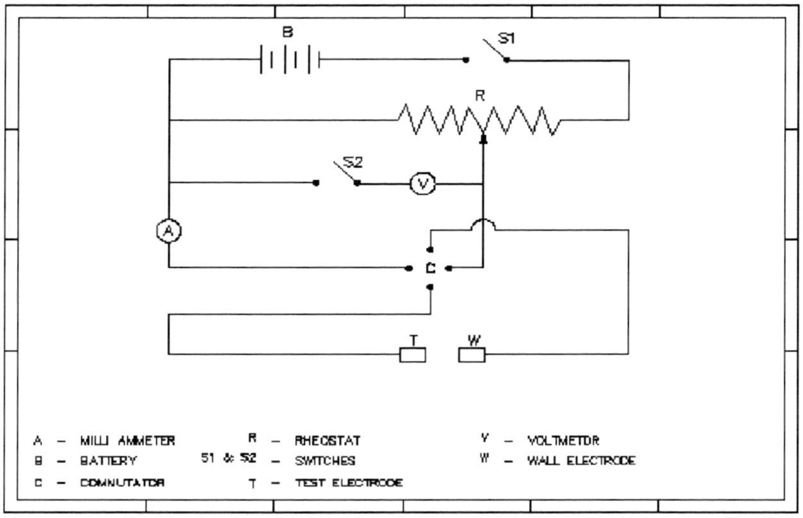

Figure 4: Circuit Diagram

RESULTS AND DISCUSSION

Perforated disc is taken as turbulence promoter in the present study. The disc is having four round holes in the four quarters. Each perforation is of 2mm.The disc is made up of Nylon. Any turbulence promoter enhances mass transfer rates. Transfer rates are found to vary along the test section and depend upon the region in which it falls because of localized nature of turbulence. The transfer rates are computed for each point electrode as they are placed all along the test section.

Table 2 indicates the exponent on velocity of the present study together with the other works. The exponent on velocity is comparable to that in the studies on mass transfer with different turbulence generating systems.

Table 2

Author	Promoter	System	Exponent on velocity	Range of Re
Klaczack [15]	Spiral coil	Mass transfer	0.520	1700-20000
Sujatha [16]	Tapes mounted on a rod	Mass transfer	0.490	1348-30605
Venkateswarlu [6]	String of discs	Mass transfer	0.498	3300-18650
Sitaraman [12]	String of spheres	Mass transfer	0.556	100-34000
Sarveswara Rao [13]	String of cones	Mass transfer	0.431	690-20200
Changal Raju [14]	Wires wound on a rod	Mass transfer	0.490	1500-20000
Nageswara Rao [17]	Tape-disc assembly	Mass transfer	0.485	1300-12000
Teja Latha et al [18]	Square grooved serrated disc	Mass transfer	0.205	1933-19337
Present study	Perforated disc	Mass transfer	0.488	5264-21059

Effect of geometric parameters:

Effect of disc diameter:

The variation of k_L with velocity (V) is drawn for different perforated disc diameters and is shown in figure 5. Various disc diameters used in the present study are d_d=0.025m, 0.030m, 0.035m, 0.040m, 0.045m.Mass transfer coefficient increases with increase in disc diameter. The augmentation in mass transfer coefficient is 1.23 times over the smooth tube [11] for disc diameter 0.025m at a velocity of 0.3936m/s while the augmentation is 1.79 times over the smooth tube [11] for the disc diameter 0.045m at the same velocity 0.3936m/s.

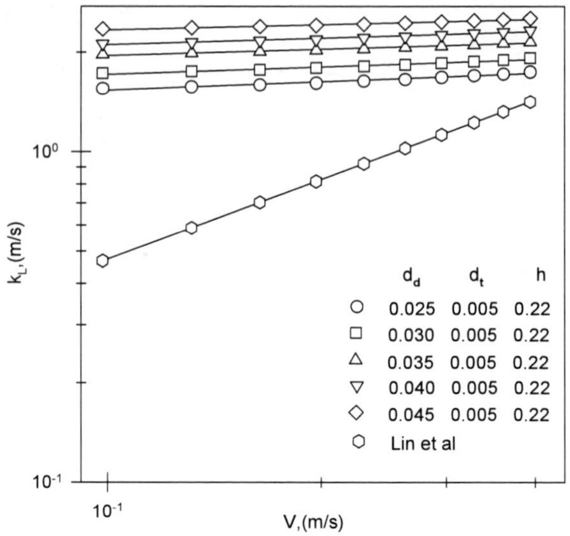

Figure 5: Variation of mass transfer coefficient with velocity - Effect of diameter of perforated disc

Effect of disc thickness:

In figure 6, mass transfer coefficient K_L is drawn against velocity to study the effect of disc thickness on mass transfer by keeping all the other parameters constant. The thickness of the disc used in the present study are d_t=0.005m, 0.020m, 0.040m, 0.060m. Mass transfer coefficient increases from 1.79 times to 5.95 times as the thickness of the disc increases from 0.005m to 0.060m at the velocity of 0.3936 m/s.

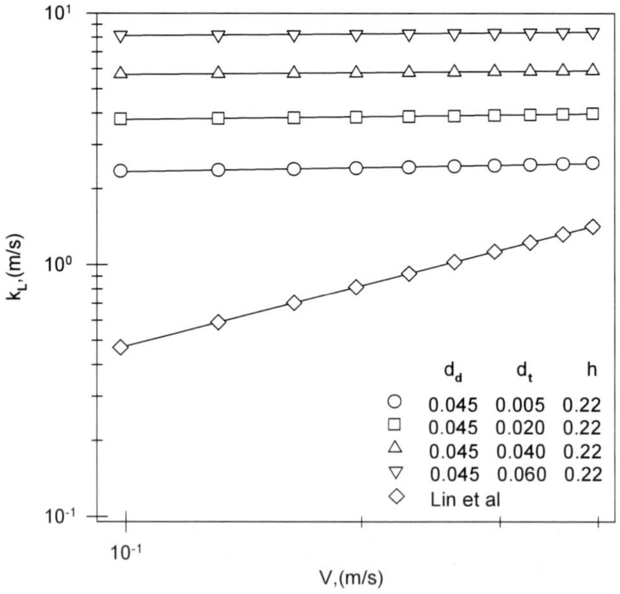

Figure 6: Variation of mass transfer coefficient with velocity -Effect of disc thickness

Effect of spacing inside the

test section:

As the distance of the perforated disc from the entrance of the test section varied, extent of turbulence also varied because of change in circulating pattern and it is extended to both perforated disc region and down stream region to disc region. Mass transfer coefficient (K_L) versus velocity for different distances of the disc(h) from the entrance of the test section are plotted and is shown in figure 7. Mass transfer coefficients are decreased from 2.09 to 1.5 times over Lin et al [11] for the smooth tube, while the distance of the disc from the entrance of the test section increases from 0.14 m to 0.30m.

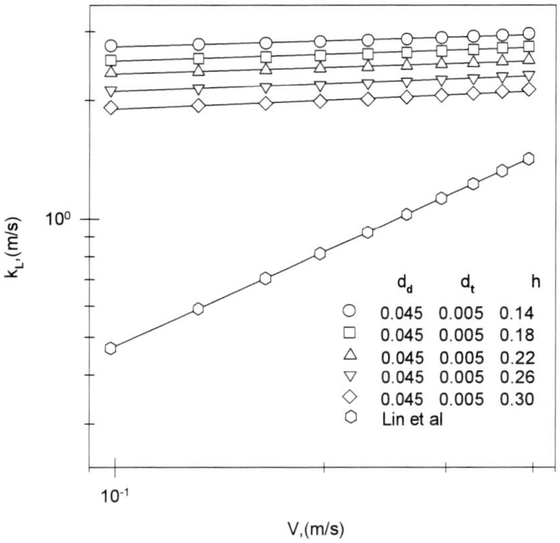

Figure 7: Variation of mass transfer coefficient with velocity -Effect of distance of disc from entrance of test section

DEVELOPMENT OF CORRELATIONS

The data on mass transfer with single perforated disc as turbulence promoter could well be calculated in the lines done in earlier studies [17].Correlation of data using Colburn factor 'J_D' with Reynolds number have yielded the following equations.

Perforated disc region:

$J_D = 83.21\ (Re)^{-0.9348}$ ....(3)

Average deviation = 44.35

Standard deviation = 68.98

By incorporating dimensionless geometrical groups, the following correlations are yielded

Perforated disc region:

$J_D = 0.155(Re)^{-0.9432}(\phi_1)^{0.641}(\phi_2)^{0.4658}(\phi_3)^{-0.3775}(Sc)^{1.1521}$ ....(4)

Average deviation = 3.53

Standard deviation = 5.25

Where $\phi_1=d_d/d$, $\phi_2=d_t/d$, $\phi_3=h/d$, S_c which are dimensionless groups and Sc is Schmidt number

Correlation plot for equation (4) is presented in the figure 8.

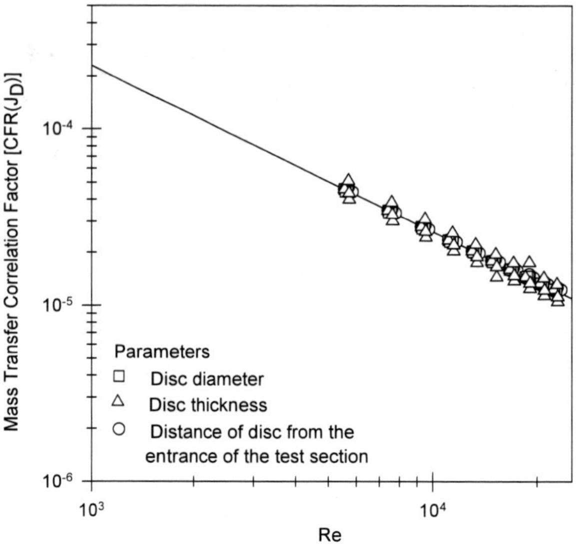

Figure 8: Correlation plot for equation 4

Comparison of correlations:

For a selected set of geometric parameters correlation factor in mass transfer (Y_3) against Re is plotted, for comparison data of other studies namely Tejalatha [18] having comparable geometric parameters, are computed with present method is shown and plotted as figure 9. The data falls close to the present study indicating correlation presented in the present work is comparable.

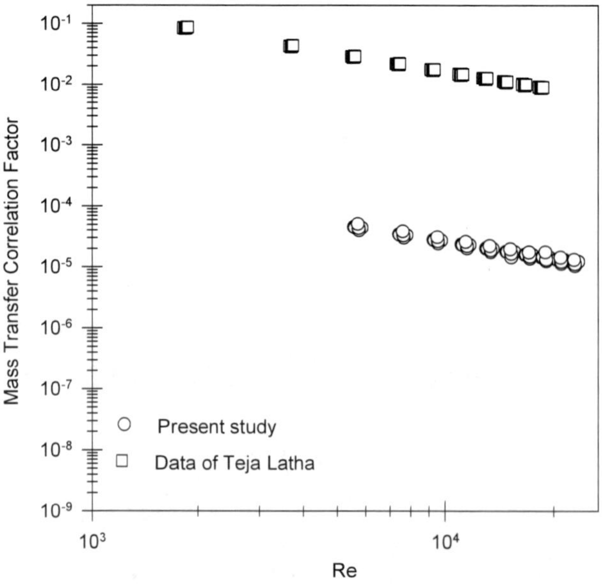

Figure 9: Variation of CFR with Re
- Comparison plot

CONCLUSIONS

Mass transfer coefficients are increasing with increase in velocity. Mass transfer coefficients are increasing with increase in disc diameter (d_d), thickness of the disc (d_t) and decreases as distance of the disc (h) from the entrance of the test section increases. In the present study, it is found that disc diameter (d_d) of 0.045m, disc thickness=0.005, distance of the disc from the entrance of the test section (h) =0.14m, offered maximum augmentation. A maximum augmentation of 5.95 folds is observed over smooth tube flow. Correlations developed based on semi theoretical considerations. Wall similarity concept is applied for the present case.

Correlations developed for mass transfer:

Perforated disc region: $J_D = 0.155(Re)^{-0.9432}(\phi_1)^{0.641}(\phi_2)^{0.4658}(\phi_3)^{-0.3775}(Sc)^{1.1521}$

Dimensionless Groups:

J_D = Mass Transfer Factor $(k_L/V).S_C^{2/3}$

Re = Reynolds number = $dV\rho/\mu$

Re^+ = Roughness Reynolds number = $(d_d / d).Re.\sqrt{(f/2)}$

$R(h^+)$ = Roughness momentum transfer function = $2.5\ln(2\, d_d/d) + \sqrt{(2/f)} + 3.75$

St = Stanton number = k_L/V

St_o = Stanton number for conduit without internals

Sc = Schmidt number $\mu/\rho D_L$

Sh = Sherwood number $k_L d/D_L$

u^+ = Dimensionless velocity, u/u^*

y^+ = Dimensionless radial distance from the wall, $y u^*/\nu$

Nomenclature:

d = Diameter of test section, m

D_L = Diffusivity of reacting ion, m^2/sec

d_d = Disc diameter, m

E = Energy consumed using perforated disc in the conduit, N/m^2

E_o = Energy consumed for empty conduit, N/m^2

f = Friction factor, $\Delta p\, d\, g_c/2LV^2 \rho$

ΔP = Pressure difference, N/m^2

F = Faraday's constant = 96,500 coulombs/g-mol

g = Acceleration due to gravity, m/sec^2

g_c = Gravitational constant.

i_L = Limiting current, amp

k_L = Mass Transfer coefficient, m/s

k_o = Mass transfer coefficient of the empty conduit, m/s

L = Length of Test section, m

n = Number of electrons transferred

Q = Volumetric flow rate, m^3/s

d_d = Diameter of the disc, m

d_t = Thickness of the disc, m

h = Location of the disc from the entrance of the test section, m

u = Local velocity, m/s

u^* = Friction velocity = $\sqrt{(\tau_w g_c / \rho)}$, m/s

V = Average velocity, m/s

y = Radial distance from the wall, m

Y_3 = $J_D / (d_d/d)^{0.641} (d_t/d)^{0.4658} (h/d)^{-0.37750} (Sc)^{1.1521}$

REFERENCES

[1]. Burns, J.R. and Jchuck,R.J.J., "Determination of liquid – solid mass transfer coefficient for a spinning disc reactor using limiting current technique", Int. J. Heat and Mass Transfer, 2005, 48, 2540-2547

[2]. Ravi,T. , Srinivasa Rao, B. Gopal Kishna, P. and Venkateswarlu,P., "Ionic mass transfer studies in fluidized beds with co-axially placed discs on a rod as internal", Chem. Engg . Process., 1996,35, 187-193

[3]. Sherwood, J.D., and Stone, H.A., "Added mass of a disc accelerating within a pipe, Phy. Fluids., 1997, 9,pp 3141-3148.

[4]. Padesta, J.J., Paus, G.S., and Arvia, A.J., "Ionic mass transfer at horizontal disc electrode under longitudinal vibration", Electro Chem. Acta , 1974,583-589.

[5]. Escudie, R., Epstein, N., Grace, R., and Bi, H.T., "Effect of partical shape on liquid fluidized beds of binary solids mixtures", Chem.Engg. Science., 2006, 61,1528-1539 .

[6]. Venkateswarlu, P., Gopichand, T., and Raju, G.J.V.J., "Increased mass transfer in a circular column in the presence disc promoter", J.Energy Heat Mass Transfer, 2000, 22,195-203

[7]. Krishna, M.S., Jagannadha Raju, G.J.V. and Rao, C.V., Ind. J. Technol., 1966, 4, 8

[8]. Sudhakhara Rao, K., Rama Raju, C.V. and Jagannadha Raju, G.J.V., Ind. J. Technol.,1966, 6, 129

[9]. Bhaskar Sarma, C., "Ionic mass transfer at circular and elliptical cylinders in cross flow of homogeneous fluid and in fluidized solids", Ph.D. Thesis, Andhra University, Visakhapatnam 1978

[10] Venkateswarlu, P., "Studies on ionic mass transfer with co axially placed discs on a rod as turbulence promoter", Ph.D., Thesis, Andhra University, Visakhapatnam, 1987.

[11] Lin C.S.,E.B.Denton,H.S.Gaskill and G.L.Putnam, "Diffusion controlled reaction", Ind.Engg.Chemistry, 1951, 43, 2136-2143 .
[12] Sitaraman, T.S, "Augmentation of mass transfer with coaxial string of spheres as internals in tubes and fluidized beds", 1977, Ph.D. Thesis, University of Madras, India
[13] S.Sarveswara Rao, G.J.V.J.Raju, "Studies on Ionic Mass Transfer with co-axially placed cones on a rod in homogeneous fluid and fluidized beds", Paper presented at World Congress III of Chemical Engineers, 1986,Tokyo, p481.
[14] D.Changal Raju,C.Bhaskara Sharma,G.J.V.Raju, "Ionic mass transfer at the outer wall of concentric annuli in the presence of fluidizing solids-Effect of wires wound on central rod", J.Energy Heat Masss Transfer, 1992, 14,151-156.
[15] Klaczac,A, Trans. A.S.M.E., J. Heat Transfer, 1973, 95,557
[16] V.Sujatha,C.Bhaskara Sarma,G.J.V.Jagannadha Raju, "Studies on ionic mass transfer with coaxially placed helical tapes on a rod in forced convection flow", Chem. Engg. Process.,1997,36, 67-73
[17] Nageswara Rao, V and Chitti Babu, N, "Studies on ionic mass transfer with coaxially placed disc as turbulence promoter in circular conduits", Caledonian journal of Engineering, 2009, Vol.4.
[18] S. Teja Latha, V.Nageswara Rao and M.Gangadhar, "Mass transfer in circular tube with square-grooved serrated disc as turbulence Promoter", Int. J. Chemical Engineering Research, 2010, vol. 02, pp.335-344.

(117d) Process Improvement in the Energy Recovery Ammonia Synthesis Reactor of a Solar Thermal Power Plant

Zafar Koreshiand Sadaf Siddiq, Mechatronics Engineering, Air University, Islamabad, Pakistan

Abstract—Among renewable energy sources, solar energy has been harnessed from photovoltaic (PV) technologies as well as concentrated solar power (CSP) technologies, both with the drawback of not being able to operate as baseload plants in the same manner as conventional fossil-fueled or nuclear power plants. Thus, thermal storage during solar insolence is necessary to permit full 24-hour operation of a power plant. Among the storage options, liquid ammonia, a chemical available in abundance, has been investigated as a thermochemical storage medium. During solar insolence, the solar thermal energy is used for the endothermic cracking of ammonia to nitrogen and hydrogen; and recovered subsequently by exothermic synthesis in the well-known industrial Haber-Bosch Process.

This paper considers optimal process parameters of two power plants, 1 kW to 10 MWe, in which an ammonia loop is used for energy recovery during periods when solar insolence is not available. Two industrial designs***viz***a Kellogg Brown Root (KBR) horizontal reactor and a Haldor Topsøe type vertical reactor, are used in the analysis.

The models are based on the one-dimensional mass and energy conservation equations with the thermodynamics and reaction kinetics expressed by the Temkin-Pyzhev relation. The optimal requirements are obtained from a variational formulation seeking to maximize the heat of reaction from the synthesis of ammonia. A numerical solution is carried out to quantify the effect of system pressure, spatial catalyst concentration, and activation energy on the energy recovery system. The results indicate that significant process improvements, such as reduction in system pressure and close-to-equilibrium operation, are possible by varying the spatial catalyst distribution to take advantage of the 'importance' of the first catalyst bed in a series of beds. An 'optimal' catalyst loading in a reactor is shown to yield the 'best' temperature profile resulting in maximum ammonia conversion. The optimal distributions are compared with equilibrium, temperature and concentration, profiles. It is found that a close-to-equilibrium conversion can be achieved by placing a high-activity catalyst in the first bed resulting in enhanced heat recovery when high-activity, or spatially concentrated, catalysts are placed in 'important' zones of the synthesis reactor.

(117e) Heat Integrated Reactive Distillation of Mesityl Oxide Process

John C. Prindle[1],**C. B. Panchal**[2], Richard D. Doctor[3], Emmanuel A. Dada[4], Aspi Kolah[5], Carl T. Lira[6]and Dennis Miller[6], (1)Argonne National Laboratory, Argonne, IL, (2)Argonne National Laboratory, Chicago, IL, (3)Energy Systems Division, Argonne National Laboratory, Argonne, IL, (4)ChemProcess Technologies (CPT), LLC, League City, TX, (5)Michigan State University, East Lansing, MI, (6)Chemical Engineering and Materials Science, Michigan State University, East Lansing, MI

Significant advances have been made by industry and academic research organizations to establish the commercial advantages of Reactive Distillation (RD). It is an effective but challenging technique for energy-efficient synthesis of major commodity chemicals with added advantages of process simplification. However, the following key technical barriers hinder wider application of RD: a) effective ways to improve contact between the catalyst and reacting phases; b) effective thermal management approaches to maintain an optimal global temperature profile that would also influence global distribution of reactants and the localized temperature distribution in the catalyst bed; and c) a fundamental rate-based model to elucidate the effect improvements in thermal management and catalyst contact might have on the localized reactive multi-phase flows. This presentation will focus on the potential benefits of implementing several different thermal management approaches to a reactive distillation system. The production of diacetone alcohol (DAA) and mesityl oxide (MO) from acetone dimerization and subsequent dehydration is used for illustration. This reaction system is a good candidate for RD and experimental results have been previously reported in the literature. Current literature analysis shows that the selectivity for DAA is quite sensitive to the reboiler heat duty, which indicates the thermal management in this RD process can play key role in optimizing the product selectivity, conversion rate, and energy efficiency. In this analysis an Aspen+ model for a heat integrated reactive distillation (HIRD) was developed. In one configuration, the latent heat from the overhead condenser was recovered and introduced at selected stages to determine the optimum heat integration. Other configurations evaluated involved both the addition and removal of heat at different stages. While the primary focus of the presentation will be on configurations with the greatest energy savings at constant throughput and purity, the effect of HIRD on production rate at a constant energy requirement will also be discussed.

(143a) Cationic Polyelectrolytes Decrease Gelatinization Temperature of Cornstarch

Sandeep Mora, Chemical and Biomolecular Engineering, Georgia Institute of Technology, Atlanta, GA, Kendra E. Maxwell, School of Chemical & Biomolecular Engineering Georgia Tech, Georgia Institute of Technology, Atlanta, GA and Sujit Banerjee, School of Chemical and Biomolecular Engineering, Georgia Institute of Technology, Atlanta, GA

Cationic polyelectrolytes such as c-PAM and poly-DADMAC decrease the gelatinization temperature of cornstarch granules. Enzymatic hydrolysis of cornstarch is, therefore, more efficient in the presence of the polymers. Viscosity profiles of various combinations of the cornstarch/enzyme/polymer system will be presented as will corresponding glucose and Brix values. A mechanism for the effect of the polymer will be proposed and discussed with reference to previous work on polyelectrolyte assisted hydrolysis of bleached cellulosic fiber.

(143b) Cloud Point Studies for Diesel and Jet Fuel Bio-Derived Fuels

Anne Lown, Lars Peereboom, Dennis J. Miller and Carl T. Lira, Chemical Engineering and Materials Science, Michigan State University, East Lansing, MI

Alternatives to petroleum based fuels are essential for the future transportation industry. For a fuel to be a viable replacement for a petroleum fuel it must meet a number of specifications. This work considers the compatibility of potential biofuels with diesel and jet fuel. This research focuses on low temperature properties.

Cloud point measurements are presented for additives in two test diesel blends and jet fuel. Additive functionalities studied include, diesters, esters, ketones, ethers, and alkanes. Diesters, as a class, appear to create liquid-liquid immiscibility and a cloud point higher than that of either the base fuel or the pure diester. Esters decrease the cloud point of diesels fuels, as do most ketones. Ethers and alkanes respectively decrease the cloud point even more. These results are consistent with the precipitation of a component of the petroleum fuel, rather than an additive, and the decrease in cloud point is due to a dilution effect. For some ketones, a different effect is observed. Based on the pure component melting temperature of the ketone, the compounds studied cause decreases or increases.

Long chain ester components were also tested. These additives have higher cetane numbers than other potential fuel components, and are therefore more desirable as a diesel fuel component. Ketones and ethers of similar lengths were also tested, and have the same behaviors as described above. The behavior of potential fuel additives seems to be more related to the oxygen containing functional group rather than extent of branching or size.

Experimental results will be interpreted with thermodynamic models using modified surrogates.

(143c) Process Development and Demonstration of Biomass to Liquid (BTL) Fuels Via Gasification and Catalytic Conversion

Qiangu Yan, Department of Agricultural & Biological Engineering, Mississippi State University, Mississippi State University, MS; Dave C. Swalm School of Chemical Engineering, Mississippi State University, Mississippi State University, MS and Fei Yu, Department of Agricultural & Biological Engineering, Mississippi State University, Mississippi State, MS

Biomass like wood chips, switchgrass and other plant residues are first converted to syngas through gasification process using air, oxygen or steam. A downdraft gasifier is performed for syngas production in Mississippi State. The syngas from the gasifier contains about 20% H_2, 20% CO, 11% CO_2, and 49% N_2. In addition to the gases listed above, the bio-syngas also contains some tars, ammonia, H_2S, and particulates as impurities. Bio-syngas has to be polished prior to CO hydrogenation to liquid fuels. The purpose of the work is to catalytic upgrade cleaning syngas to higher hydrocarbons and value-added chemicals. A series of catalysts have been developed to produce gasoline, diesel, jet fuels, olefins, higher alcohols and esters. Current work successfully demonstrates the process of biomass to liquid (BTL) fuels via gasification and catalytic conversion.

(143d) Global Optimization of Oil Production in Microalgae Bioreactor

Javad Abdollahiand Stevan Dubljevic, Chemical and Materials Engineering, University of Alberta, Edmonton, AB, Canada

Sustainability and renewability of bio-fuels make them suitable replacements for fossil fuels, however because of slow production rate and high production costs, theyare not economically comparable with traditional fuel sources. Bio-fuel production fromphototrophic and heterotrophic microalgae is a novel method of oil production that isgoing to replace bio-fuel production from crops. In this method there is no need forvast land, high amounts of water and in the case of heterotrophic microalgae there is noneed for sunlight. Furthermore, increasing oil production rate can ensure the economicfeasibility of biodiesel production.

This work is built on an existing heterotrophic microalgae fed-batch model, see [1], whichdescribes the algae growth and oil production. The fed-batch microalgae reactor is fedby two nutrients, glucose as carbon source and glycine as nitrogen source. In this work,maximum possible oil production in the shortest time interval is calculatedby using a global optimization technique. The highly nonlinear rate functions, timevarying parameters and uncertainties in the modeling are taken into accountwithin the optimization framework.

Reference:

[1] H. De la Hoz Siegler, A. Ben-Zvi, R.E. Burrell, W.C. McCa rey, The dynamics of heterotrophicalgal cultures, Bioresource Technology 102 (2011) 5764-5774

Sustainable Biodiesel Production from Waste Cooking Oil and Animal Fats under Supercritical Methanol

AICHE 2012 Spring Meeting, Houston, Texas

Chee Kai Tan, Maoqi Feng, Laura Harris, and Cary Wong
Chemical Engineering Department, Chemistry and Chemical Engineering Division, Southwest Research Institute, San Antonio, Texas

Abstract

In the conventional process for transesterification of vegetable oils and animal fats to biodiesel, free fatty acids and water always present a problem and result in a low conversion. In this study, triglycerides, animal fats, and waste cooking oil were used as feedstock for transesterification under supercritical methanol conditions. For glyceride tributyrate, the conversion was 100% under SCM conditions. For waste cooking oil, the fatty acid methyl ester (FAME) yield was >90% at the following reaction conditions: reaction temperature = 300°C and 250°C, reaction pressures = 2000 psi, 2500 psi, and 3000 psi, residence time = 18 minutes, and molar ratio of methanol to oil = 40:1. However, when the residence time decreased to 7 minutes, at 250 °C and 2500 psi, the transesterifcation conversion dropped to 86 %. Using a heterogeneous catalyst, ETS-10, the transesterification reaction conditions were lowered to subcritical levels, and a conversion of over 99% was achieved with a residence time of 7 minutes. Because of the short residence times, reduced waste, and high triglyceride conversions, a biodiesel process employing supercritical methanol may present significant economic advantages over the conventional manufacturing process.

Introduction

As the availability of petroleum-derived diesel fuel slowly decreases, the interest in alternative fuels grows. Biodiesel refers to a vegetable oil or animal fat based fuel consisting of long-chain fatty acid alkyl esters (FAMEs). It is a renewable alternative fuel and has been successfully derived from lipids (vegetable oil or animal fat) and an alcohol as shown in Equation 1.

$$
\begin{array}{llllllll}
CH_2\text{-OCO-R} & & & & & & CH_2\text{-OH} & \\
| & & & & & & | & \\
CH\text{-OCO-R} & + & 3R'OH & \longrightarrow & 3RCOOR' & + & CH\text{-OH} & (1) \\
| & & \text{Alcohol} & & \text{Alkyl Ester} & & | & \\
CH_2\text{-OCO-R} & & & & & & CH_2\text{-OH} & \\
\text{Triglyceride} & & & & & & \text{Glycerol (byproduct)} &
\end{array}
$$

R = fatty acid chain
R' = CH_3 : fatty acid methyl esters
R' = CH_2CH_3 : fatty acid ethyl esters

Biodiesel is becoming well-known as an environmentally friendly fuel due to its nontoxic and biodegradable characteristics. It has reduced engine emissions of sulfur oxides, carbon dioxide, particulate matter, and aromatic hydrocarbons compared to conventional diesel fuels. Due to federal legislation to promote energy independence (EISA) and the consequent mandates issued by the EPA (in RFS2) requiring the increased production of transportation biofuels, biodiesel has received considerable attention: it appears to be one of the most promising of the biofuels for meeting the production volume requirements.

Research into the use of transesterified sunflower oil as diesel fuel was initiated in South Africa in 1979. By 1983, the process for producing fuel-quality, engine-tested biodiesel was completed and published internationally [1]. Biodiesel is commercially produced in batch or continuous stirred tank reactors at temperatures ranging from 60 to 200°C using homogeneous base catalysts with edible vegetable oils such as palm oil, corn oil, sunflower oil, and soybean oil. The fatty acid triglycerides are converted to the respective alkyl ester (biodiesel) and glycerol by acid-catalyzed [2-4] or base-catalyzed [5-11] transesterification in the presence of an excess of methanol or ethanol.

However, the use of homogeneous base catalysts requires neutralization and separation from the final reaction mixture and results in higher solvent consumption. Thus, these additional production costs make the product biodiesel not cost competitive with fossil diesel. Heterogeneous solid base catalysts, able to catalyze the transesterification of triglycerides to alkyl eaters, could solve these problems. The solid catalysts can be separated from the biodiesel without using solvent and they can be easily regenerated [12-17].

Transesterification of oils to alkyl esters in supercritical methanol without a catalyst was developed by Saka and Kusdiana [18-20] and offered a promising future for industrial applications. Supercritical fluids are fluids above their critical temperature and pressure. The supercritical fluid behaves gas-like considering diffusivities but also liquid-like with respect to viscosities. As shown in Figure 1, the critical temperature and critical pressure of methanol are 240°C and 1140 psia, respectively [21].

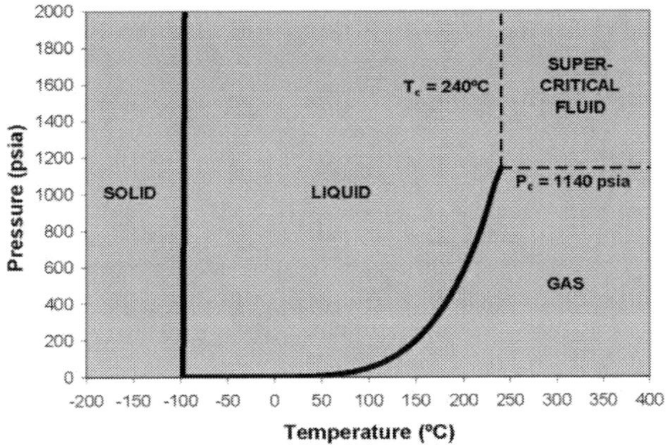

Figure 1. Supercritical methanol phase diagram.

Nonpolar triglycerides can be well solvated with supercritical methanol to form a single phase oil/alcohol mixture. In addition, the conversion rate of oil into esters increases dramatically in supercritical alcohols, the reaction does not require a catalyst, and it is not affected in the presence of water and free fatty acids [22-28]. In a supercritical alcohol, the transesterification reaction was nearly completed in less than 15 minutes for both edible and non-edible oils [26]. Using the supercritical methanol method for conversion waste cooking oil at 1450 psi and a temperature of 300°C, the yield of biodiesel was shown to increase to 50-65% in the first 15 minutes [27]. Recently, a continuous production process for vegetable oil conversion to biodiesel through the supercritical methanol (SCM) has been reported [29-31]. However, continuous SCM processing of waste cooking oil and animal fats has not been studied. This paper summarized our study of such a process.

Experimental Section

1) Reactor System: A flowsheet of the continuous transesterification fixed-bed reactor system with supercritical methanol is shown in Figure 2. Figure 3 [32-36] is a photo of the system as built in the Chemical Engineering Center at Southwest Research Institute.

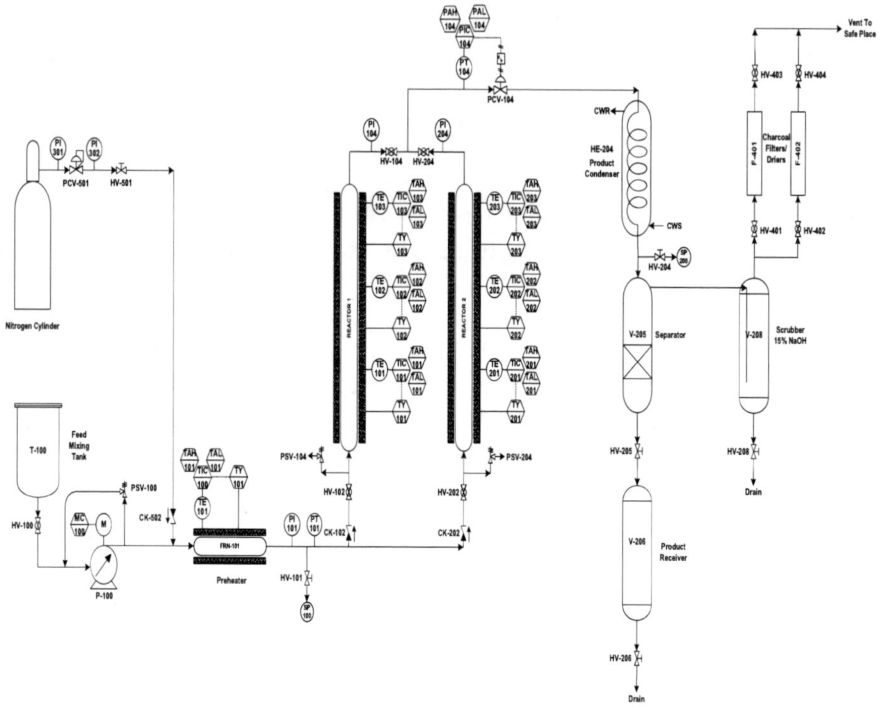

Figure 2. Process Flow Diagram for biodiesel production with supercritical methanol.

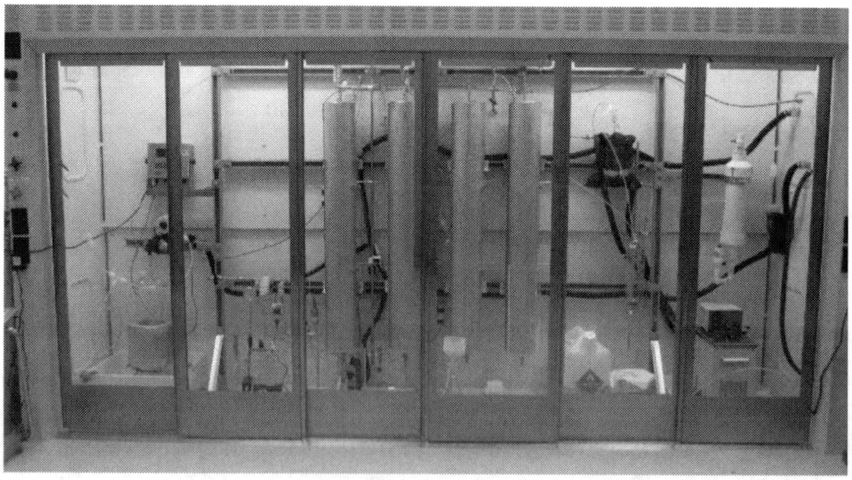

Figure 3. Continuous transesterification bench unit at Southwest Research Institute.

The reactor was constructed from a 3/8" OD x 4' long stainless steel tube with a wall thickness of 0.049 inch, giving a volume of 47.4 mL. The reactor was built with two thermowells welded into the reactor tube and located one third of the way from each end of the reactor. Thermocouples inserted into the thermowells monitored the internal temperature of the chemicals as well as the catalyst temperature. The stainless steel reactor was wrapped with heat tape and installed inside an aluminum enclosure insulated with a calcium silicate material.

The pressure within the reactor system was recorded with a pressure transducer and regulated with a pressure control valve, both located downstream of the reactor. At the outlet of the pressure controller, the biofuels, the bio-based byproduct, glycerin, and excess methanol were then fed into a condenser and collected in a graduated glass receiver. Uncondensed gases were captured within an activated-carbon scrubber. For safety, pressure relief valves were installed along the reactor and the transfer lines. The process instrumentation was automatically controlled with an i-Fix program. The software allowed for data trending, automatic controls, safety alarms and fail-safe process shutdown.

A 4L feed preparation vessel mixed the reagents, feedstock and methanol, using as overhead mechanical stirrer. The feed vessel was located on a weigh scale so that the delivery of reagents to the reactor could be measured for material balances. The reagents were pumped to the reactor using a high-pressure Eldex metering pump (Model B-100-S, flow rate from 0.2 – 8.0 mL/min @5000 psig). The inlet of the reactor was equipped with a pressure gauge and transducer to monitor the liquid feed pressure. A pressure safety valve was installed at the inlet of the reactor and set at 4,000 psig.

2) Feedstock Preparation

Three feedstocks were prepared: methanol/tributyrin (40:1 molar ratio), methanol/waste cooking oil (40:1 molar ratio), and methanol/pork fat (>200:1 molar ratio). The first feedstock, consisting of methanol and tributyrin, was a single phase, clear solution. No filtration or other purification was needed before combining the compounds or running them through the system.

The waste cooking oil (WCO) was filtered first through a 10-micron filter followed by a 0.7-micron filter to remove suspended impurities that could potentially plug the system, particularly the catalyst pores. After combining this filtered WCO with the appropriate amount of methanol based on stoichiometry, the solution was biphasic and thus had to be constantly mixed to maintain a homogeneous consistency while being pumped into the system.

The pork fat was in a semisolid state before use, and it therefore had to be heated prior to filtering. The fat was filtered through a 10-micron filter before it was combined with methanol in a 200:1 molar ratio of methanol to pork fat. This molar ratio was increased from the typical 40:1 molar ratio in order to solubilize more of the pork fat in the methanol. The feedstock was biphasic and was mixed constantly.

3) Catalysts Preparation

Microporous Titanosilicate ETS-10 Catalyst [37]

184.1 g of water and 182.5 g of sodium silicate solution (10.6 wt % Na2O, 26.5 wt % SiO_2) were added to a Teflon beaker and the mixture was stirred. While stirring, 63.2 g NaCl, 12.3 g KF, and 12.3 g TiO_2 were added into the silicate solution producing a translucent thick gel. After the solution was stirred vigorously for 30 minutes at room temperature, the mixture contained in the Teflon beaker was placed into a Parr autoclave and heated to 200 ^{o}C for 42 hours. The gel had the composition $5.3SiO_2$:$1.0TiO_2$:$1.6Na_2O$:7.2NaCl:1.0KF or KCl:$110H_2O$, and the pH was around 10.5. The samples obtained were denoted as microporous titanosilicate ETS-10.

4) Analytical Methods

The biodiesel was analyzed with a Hewlett Packard GC Model HP 6890 interfaced with an Agilent 5973 Network Mass Selective Detector (Figure 9). The analytical column was a DB-23 fused-silica capillary column (0.25mm ID x 60m length) with a 0.15μm film thickness silicone-coated fused-silica capillary column. This narrow-bore capillary column was directly connected into the mass spectrometer inlet.

Direct measurement of triglycerides present in the converted product samples were accomplished using high performance liquid chromatography (HPLC) coupled with an evaporative light scattering detector

(ELSD). By analyzing feedstock and product samples, a triglyceride conversion rate was determined. The analytical method was developed on a Waters HPLC system model LC03, utilizing a 600E pump, 717 autosampler and an Alltech 2000 ELSD. A Supleco brand analytical column (part number Supelcosil LC-18-DB, 150 mm x 4.6 mm.) was chosen for the resolution of the triglycerides present in the samples.

Results and Discussion

1) Baseline: Transesterification of tributyrin under supercritical methanol conditions

Tributyrin was used as benchmark compound to guide the experiment. Table 1 shows the results of the GC/MS analyses conducted on the tributyrin samples. The data indicate that at temperature below supercritical condition of methanol (SCM), there was no conversion of tributyrin. At temperature above SCM (240 °C, 1140 psi), tributyrin was completely transesterified to butyric acid methyl ester with methanol.

Table 1. **GC-MS Results and Percent Conversions**

Run ID:	6A	6B	6C	6D	6E	6F
Feedstock	Tributyrin	Tributyrin	Tributyrin	Tributyrin	Tributyrin	Tributyrin
Pressure Stpt (psig)	2000	2500	3000	2000	2500	3000
Temp Stpt (°C)	200	200	200	300	300	300
Initial Tributyrin Concentration (mg/L)	158147	158147	158147	158147	158147	158147
Final Tributyrin Concentration (mg/L)	169000	180000	150000	0	0	0
Percent Conversion (%)	-6.86	-13.82	5.16	100	100	100

2) Transesterification of WCO under supercritical methanol conditions

The results of transesterification of WCO with SCM (> 240 °C/1140 psig) are tabulated in Table 2. The results show that high conversions of triglyceride into fatty acid methyl esters (FAMEs) with SCM are consistently achievable on a continuous basis with reaction conditions in the supercritical region for methanol: temperature ranges 250 – 325 °C and pressure ranges 1500 – 3000 psig at a feedstock flow rate of 2.5 mL/min.

Table 2. Waste Cooking Oil (WCO) Biodiesel Conversion with Supercritical Methanol (SCM).

Sample ID	Temp. °C	Pressure, psig	Flowrate, mL/min.	Residence Time, min.	TG (mg/mL) Concentration	Triglyceride Conversion (%)
R8145-8A-033	300	1500	2.78	17.16	33	**88.2**
R8145-8B-035	300	2500	2.6	18.35	0.34	**99.9**
R8145-9A-037	250	1500	2.53	18.85	7	**97.5**
R8145-9B-039	250	2500	2.2	21.68	0.93	**99.7**
R8145-9C-041	250	3000	2.4	19.88	0.56	**99.8**
R8145-9D-043	300	3000	2.58	18.49	0.56	**99.8**
R8145-10A-045	325	1500	2.3	20.74	0.54	**99.8**
R8145-10B-046	325	2500	2.55	18.71	0.58	**99.8**
R8145-11A-049	325	1500	2.4	19.88	8.2	**97.1**
R8145-11B-050	325	2500	2.67	17.87	0.59	**99.8**
R8145-25A-076	250	2500	4.7	10.15	17	**93.9**
R8145-25B-077	250	2500	5.57	8.56	26	**90.7**
R8145-25C-078	250	2500	7	6.81	38	**86.4**

R8145-28A-083	250	2000	2.83	16.86	3.6	**98.7**
R8145-28B-084	250	1700	2.86	16.68	2.2	**99.2**
R8145-28C-085	220	2200	3	15.9	18	**93.6**

However, as shown in figure 4, the conversion dropped to 86.4% when the flow rate increased from 2.2 mL/min (residence time 21.68 min) to 6.81 mL/min (residence time 6.96 min), which indicates that triglyceride conversion is directly dependent on the contact time of the triglyceride with methanol (MeOH).

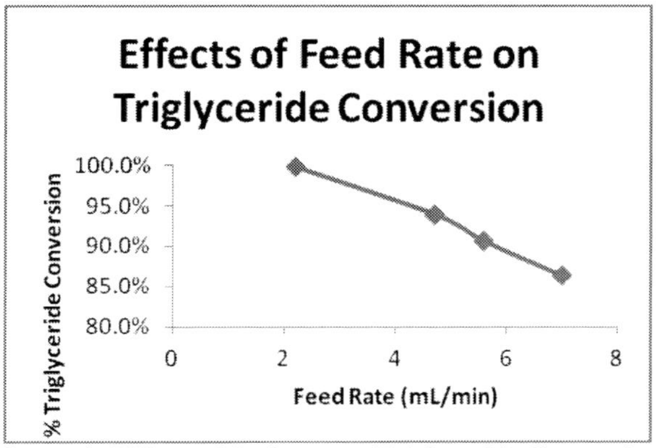

Figure 4. Effects of feed rate on Triglyceride Conversion

3) Transesterification of animal fat under supercritical methanol conditions

Table 3 shows the transesterification results of animal fat with SCM where high conversions were achieved with temperature in the range of 250 °C – 300 °C incorporated with pressure in the range of 2000 psig – 3000 psig. The Eldex pump was not able to perform with higher concentrations of pork fat due to associated plugging problems of the highly viscous feedstock.

Table 3. Pork Fat Biodiesel Conversion with Supercritical Methanol (SCM).

Sample ID	Temp. °C	Pressure, psig	Flowrate, mL/min.	Residence Time, min.	TG (mg/mL) Concentration	% Conversion
R8145-13A-051	300	2000	2.93	16.28	2	**96.4**
R8145-13B-053	300	2500	2.8	17.04	0.75	**98.6**
R8145-14A-055	300	3000	2.6	18.35	0.05	**99.9**
R8145-14B-056	250	2500	2.87	16.62	0.04	**99.9**
R8145-14C-057	250	3000	2.5	19.08	0.16	**99.7**

4) Effect of catalyst at low reaction residence times

At 250 °C/2500 psig/flowrate 6.8 mL/min., the transesterification of WCO into FAMES was low as cited in Table 2 (86.4% conversion). These reaction parameters were selected to the effect of the presence of a solid catalys by comparing the WCO transesterification activity with SCM/catalyst to SCM without catalyst. The reactor was packed with ETS-10 Catalyst and operated for twenty-four hours with SCM. Figure 5 shows that the conversion efficiency with catalytic remained active throughout 24 hours of experiment, conversion is consistently achieving >99% with a residence time of 7 minutes.

Under the same conditions, 250 °C/2500 psig/flowrate 6.8 mL/min., the transesterification of WCO to FAMEs with SCM dropped in the range of 36.5% -58.6 % in the 12 hour, nonstop experiment as shown in Figure 5. The experiment demonstrated that solid catalyst is needed at sub-supercritical conditions.

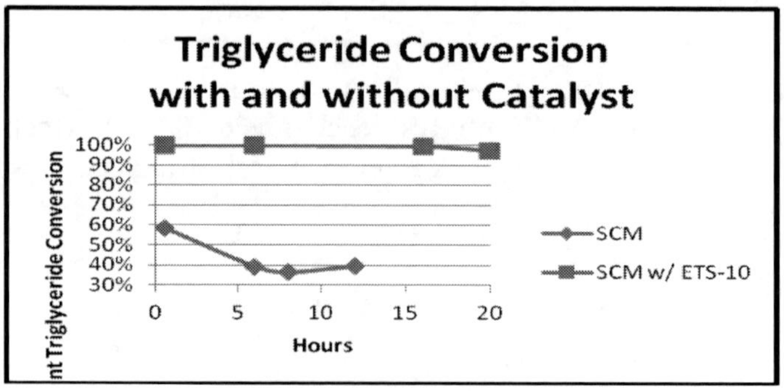

Figure 5. The effect of a solid catalyst on triglyceride conversion under SCM conditions.

Summary

Biodiesel production with continuous SCM processing of waste cooking oil and animal fats has been successfully demonstrated in lab scale in our study. Presence of a solid catalyst dramatically improves triglyceride conversion at low reaction residence times. Development of a continuous SCM biodiesel process on a pilot-plant scale is in progress at Southwest Research Institute.

Acknowledgement

The authors would like to take this opportunity to thank the Internal Research Program, Southwest Research Institute, for financial support.

References

1. K.J. Harrington and D.V. Catherine, A comparison of conventional and in-situ methods of transesterification of seed oil from a series of sunflower cultivars, J. Am. Oil Chem. Soc. 62 (1985), pp. 1009-1013.
2. J. Otera, Transesterification, Chem. Rev. 93 (1993), pp. 1449-1970.
3. B. Davies and G.V. Jeffreys, Continuous transesterification of ethyl alcohol and butyl acetate in a sieve plate column: II. Batch reaction kinetic studies, Trans. Inst. Chem. Eng. 51 (1973), pp. 271-274.
4. B. Freedman, R.O. Butterfield and E.H. Pryde, Tranesterification kinetics of soybean oil, J. Am. Oil Chem. Soc. 63(1986), pp. 1375-1380.
5. J. Schmidt, D. Reusch, K. Elgeti and R. Schomacker, Kinetics of ethanol and butyl acetate-a model system for reactive rectification, Chem Ing. Tech.71 (1999), pp. 704-708.
6. P. Felizardo, M.J.N. Correia, I. Raposo, J.F. Mendes, R. Berkemeier and J.M. Bordado, Production of biodiesel from waste frying oils, Waste Management, Vol. 26, Issue 5, 2006, pp. 487-494.
7. A.N. Phan and T.M. Phan, Biodiesel production from waste cooking oils, Fuel, Vol. 87, Issues 17-18, 2008, pp. 3490-3496.
8. S. Gryglewicz, Alkaline-earth metal compounds as alcoholysis catalysts for eater oils synthesis, Appl. Catal. A 192 (2000), pp. 23-28.

9. A. Corma, S. Iborra, S. Miquel and J. Primo, Catalysts for the production of fine chemicals, J. Catal. 173 (1998), pp. 315-321.

10. D. Damoko and M. Cheryon, Kinetics of palm oil transesterification in a batch reactor, J. Am. Oil Chem. Soc. 77 (2000), pp. 1263-1267.

11. K. Komers, J. Machek, R. Stloukal, Biodiesel from rapeseed oil and KOH 2. Composition of solution of KOH in methanol as reaction partner of oil, Eur. J. Lipid Sci. Technol. 103 (2001), pp. 359-462.

12. G. Guan, K. Kusakabe and S. Yamasaki, Tri-potassium phosphate as a solid catalyst for biodiesel production from waste cooking oil, Fuel Processing Technology, Vol. 90, Issue 4 (2009), PP. 520-524.

13. K.G. Georgogianni, A.P. Katsoulidis, P.J. Pomonis and M.G. Kontominas, Transesterfication of soybean frying oil to biodiesel using heterogeneous catalysts, Fuel Processing Technology, Vol. 90 Issue 5 (2009), pp. 671-676.

14. C. Reddy, V. Reddy, R. Oshel, and J.G. Verkade, Room-Temperature Conversion of Soybean Oil and Poultry Fat to Biodiesel Catalyzed by Nanocrystalline Calcium Oxide, Energy & Fuels 20 (2006), pp. 1310-1314.

15. C.V. McNeff, L.C. McNeff, B. Yan, D.T. Nowlan, M. Rasmussen, A.E. Gyberg, B.J. Krohn, R.L. Fedie, T.R. Hoye, A continuous catalytic system for biodiesel production, Appied Catalysis A: General 343 (2008), pp. 39-48.

16. Y.M. Park, D.W. Lee, D.K. Kim, J.S. Lee, K.Y. Lee, The heterogeneous catalyst system for the continuous conversion of free fatty acids in used vegetable oils for the production of biodiesel, Catalysis Today, 131 (2008), pp. 238-243.

17. N.S. Babu, R. Sree, P.S. Sai Prasad, and N. Lingaiah, Room temperature transesterification of edible and non-edible oils using a heterogeneous strong basic Mg/La catalyst, Energy & Fuels, 22 (2008), pp. 1965-1971.

18. D. Kusdiana, S. Saka, Effects of water on biodiesel fuel production by supercritical treatment, Bioresour. Technol. 91 (2004), pp. 289-295.

19. D. Kusdiana and S. Saka, Kinetics of transesterification in rapeseed oil to biodiesel fuels as treated in supercritical methanol, Fuel 80a (2001), pp. 693-698.

20. D. Kusdiana and S. Saka, Methyl esterification of free fatty acids of rapeseed oil as treated in supercritical methanol, Chem. Eng. Jpn., 34 b (2001), pp. 383-387.

21. R.E. Babcock, E.C. Clausen, M. Popp, W. B. Schulte, Yield Characteristics of Biodiesel Produced from chicken fat-tall oil blended feedstocks, http://ww2.mackblakwell.org/web/research/all_research-projects, 02/09/12.

22. D. Kusdiana and S. Saka, Effects of water on biodiesel fuel production by supercritical methanol treatment, Bioresource Technology, 91 (2004), pp. 289-295.

23. Y. Warabi, D. Kusdiana, S. Saka, Reactivity of triglycerides and fatty acids of rapeseed oil in supercritical alcohols, Bioresource Technol. 91 (2004), pp. 283-287.

24. W.L. Coa, H.W. Han and J.C. Zhang, Preparation of biodiesel from soybean oil using supercritical methanol and co-solvent, Fuel 84 (2005), pp. 347-351.

25. V. Rathore, G. Madras, Syntheis of biodiesel from edible and non-edible oils in supercritical alcohols and enzymatic synthesis in supercritical carbon dioxide, Fuel 86 (2007), pp. 2650-2659.

26. A. Demirbas, Biodiesel from sunflower oil in supercritical methanol with calcium oxide, Energy Conv. Manag. 48 (2007), pp. 937-941.

27. A. Isigigur, F. Karaosmonoglu, H.A. Aksoy, Methyl ester from safflower seed oil of Turkish origin as a biofuel for diesel engines, Appl. Biochem. Biotechnol., 45/46 (1994), pp. 103-12.

28. A. Demirbas, Biodiesel fuels from vegetable oils via catalytic and non-catalytic supercritical alcohol transesterifications and other methods: a survey, 44 (2003), pp. 2093-2109.

29. H. He, T. Wang, S. Zhu, Continuous production of biodiesel fuel from vegetable oil using supercritical methanol process, Fuel 86 (2007), pp. 442-447.

30. K. Bunyakiat, S. Makmee, R. Sawangkeaw, S. Ngamprasertsith, Continuous production of biodiesel via transesterfication from vegetable oils in supercritical methanol, Energy & Fuels 20 (2006), pp. 812-817.

31. H.Y. He, T. Wang, S. Zhu, Han, Continuous production of biodiesel fuel from vegetable oil using supercritical methanol process, Fuel 86 (2007), pp. 442-447.

32. Tan, C.K., Feng, M., Novel biodiesel Production Technology, National Biodiesel Technical Workshop, November 1-2, 2011, Kansas City, Kansas.

33. Tan, C.K., Feng, M., Continuous Biodiesel Production from Waste Cooking Oil and Animal Fats with Supercritical Methanol Conditions, USDA-RD State Director Conference, March 30, 2011, San Antonio, Texas.

34. Tan, C.K., Feng, M., Continuous Biodiesel Production from Waste Cooking Oil and Animal Fats under Sub/Supercritical Methanol Conditions, 2011 AIChE Spring Conference, March 13-17, Chicago.

35. Tan, C.K., Feng, M., Cooking Waste to Biofuels, USDA 9007 Renewable Energy for America Program - Workshop, January 26, 2011, Uvalde, Texas.

36. Tan, C.K., Feng, M., Transesterification of triglyceride tributyrate and waste cooking oil with supercritical methanol, 2010 ACS Winter Conference, November 30-December 4, 2010, New Orleans.

37. X. Yang, J.-L. Paillaud, H.F.W.J. van Breukelen, H. Kessler, E. Duprey, "Synthesis of microporous titanisilicxate ETS-10 with TiF4 or TiO2", Microporous and Mesoporous Materials 46 (2001) 1-11.

Unlimited Power from Continental Drift, 284207

Donald I. Garnett

Overview: Given a place to stand and a long enough lever Archimedes (1) said he could move the world. Donald I. Garnett (2, 3) observes that anchoring one continent to another with a cable will let the world's continental drift (4, 5) move you, namely: it will provide all the energy needed to power the entire worlds electrical (6) and other needs. To capture the energy of the moving continents it will be especially convenient to catenary a cable across a fault line and let the mass of the cable plus any added weights desired be lifted in elevation against gravity by the moving apart of the continents. Recovery of that energy can be accomplished using an electric generator driven by the pull of the cable in restoring the cables initial position.

Restraining the movement of a continent the size of North America in theory could generate 10^10 times the electrical usage of the United States. In principle its' restraint might stop or prevent earthquakes in California, Oregon and Washington.

The magnitude of the energy available exceeds that of tearing down the Rocky Mountains (in cable cars attached to electric motors) and dropping the rocks into the Pacific Ocean. Had the erosion by Nature of the Rocky Mountains not occurred it would be closer to equivalent but still inferior?

While this method may be applicable in many places at continental plate boundaries it is easily visualized for the Rift Valley in Africa. Places where a cable would acquire tension by continental plate movement being preferred. Movement at the face of any type fault however can be harnessed. Lateral, vertical, or oblique plate movement energy recovery can be accommodated by inclusion of a spring in the cable. Multiple echelon faults energy similarly can be harnessed using a network of spring loaded cables.

Methods: The energy associated with the drift of a continent is its ½ *mass * velocity^2 from a given anchor point. In this discussion the anchor point is another continental plate reachable by a cable. Across a fault where the movement is "apart" is especially preferred. The velocity is low but the mass is extremely large. The usual ½ mass times velocity squared is also (½*mass*velocity/410.3Kw)*velocity, or Power*velocity. With mass of a continent being say 10^22 lbs and velocity of 2.5 cm/year the power available is 10^15Kw or 10^10 times the total U. S. Electrical power.

The anchor points need not be capable of restraining the continent but may be any amount consistent with the intended power output. The small movement of the continent and the preferred relatively high velocity of the cable driving the generator begs for a velocity enhancement like a block and tackle arrangement or a geared reducer run backwards. In any event a 1000 fold block and tackle on top of another (for leverage per Archimedes) , etc., etc., will likely be in the final design. Other methods of energy recovery are possible, including:

(1) Spring compression and release,
(2) Hydraulic lifting of water for use in turbines on release.
(3) Gas compression and release.

Historically of course part of the energy of the continents movement has raised mountains and provides the force multiplier and accumulator of millions of years of potential energy generated by the continents movement. Recovery of that potential energy is available by tearing down the mountains, for example in cable cars connected to electric generators and dumping the rocks into the oceans continental trench. The Rocky Mountains, the Alps, the Andes and the Himalayas for example are available at present. Over time both energy and material (mountains) are conserved

Results: The result of using continental drift for essentially unlimited power for the human race is probably unfathomable, but nearly infinitely describable, including: No green house gasesThe hydrogen economy.

1. The elimination of nuclear power plants.
2. Electric cars with no emissions and other transportations.
3. Electric aero planes.
4. Unlimited agriculture from electric agro-lights. Production can become volumetric vice area limited, further postponing Malthus effects.
5. The largest possible scale for energy production ergo least cost.
6. Waste disposal reduced to near zero.
7. Industrial production with only carbon, water and electricity.
8. CO_2 impoundment or removal for its carbon content for global climate control.
9. Oxygen supplementation of the atmosphere for enhanced ozone protection of earth.
10. Population explosion from all of the foregoing.
11. Focused power into space.

Conclusions: Unlimited power from continental drift for the human race contains the essentials for future growth and sustainability of the race.

Some practical considerations:

While the movement of the continents has the potential power to overwhelmingly provide the needs of mankind, collecting it presents some practical problems that need to be overcome by Engineering.

Foremost among them is the issue of scale. If one supposes the power plants will be built say 100 years in the future, and the magnitude of the facility will be such as to satisfy the needs of humanity then it is safe to say the Scale will be at least 100 fold larger than the largest presently known facility. The three gorges dam and power plant in China may be a useful reference for this visualization. At 100 fold that scale the price of electricity/kwh will be down (in constant dollars) by 100 fold, assuming no increase in process complexity. Basically that means hydro electric power with a similar hydrostatic head and volumetric flow.

Major Parameters of the TGHP units and abroad power plants

Parameters	**Three Gorges Project** CHINA	**Grand Coulee** U.S.A.	**Itaipu** Brazil/Paraguay	**Guri** Venezuela	**Krasnoyarsk** Russia

Turbine					
Maximum Head(m)	113.0	108.2	126.7	146	100.5
Rated Head(m)	80.6	86.9	112.9	130	93
Minimum Head(m)	71.0(61.0)	67.0	82.9	111	76
Rated Output (MW)	710	612/716	715	610	508
Maximum Output(MW)	852	827	740/808	730	508(505)
Rated Spinning Speed(r/min)	75	85.7	90.9/92.3	112.5	93.8
Runner Diameter(m)	9.525(9.800)	9.86/9.22	8.45	7.163	7.5
Generator					
Rated Capacity(MW)	778	718	823.6/737.0	700	500
Maximum Capacity(MW)	840	710/825.6	823.6/766	805	
Frequency(Hz)	50	60	50/60	60	50
Cooling	Water Cooling of Statar	Water Cooling of Statar	Water Cooling of Statar	Air Cooling	Water Cooling of Statar
Rated Voltage(KV)	20	15	18±5%	18	15.75
Insulation Levels	F	B	F	B	
Thrust Load(t)	4,050(4,100)	4,700	4,056&4,400	2,667	
Inner Diameter of Stator(m)	18,500(18,800)	18,790	16,000	13,650	16,100

Preferred Conditions:

While this article started off with cables, multiple cables, springs, cable velocity enhancers and the passage of large amounts of time, the scale, and volumetric constraints imposed by costs lead to preferred conditions as follows and utilizes the energy accumulated over millions of years (not just a one-time event).

- Use gravity vice cables and springs.
- Use of water as the motive force for electric turbines or direct mechanical connection to the delivery conveyer. That means on the coast, close to continental trenches.
- Hydrostatic head >200 ft., so use the continental trench at 10-20,000 ft., a scale advantage of 50-100 fold.
- Transport of the mass of the continent being sent to the bottom of the trench should be as a fluid itself. Examples of two such masses are (1) the sands of the Sahara Desert, and (2) the mud volcanoes near the Mariana trench and elsewhere. An additional alternative could be molten lava, for example in Hawaii where the ocean trenches are nearby. Using
- the lava that has been cooled by Nature but is in danger of sliding off the island and creating tsunamis could be considered but I will leave that and other alternatives to others.

- This alternative suggests that the rock should be pulverized and transported as a fluid to the sea. The Western Sahara is an example of where nature over time has pulverized mountains and the wind (fluid) has transported it to the sea shore (and beyond with a gradient of about 2% out to sea). A 100 mile pipeline will yield a head for the power plant of 2 miles (10560 ft.) vs. say the 3 gorges head of 200 ft., for a scale advantage of at least 50 fold.
- The water for the electric turbines and used to transport the solid mass to the trench, may well have to be pumped from the sea to the mine, but considering the power generated by a solid of 2.5 specific gravity, dropping 10,000+ ft. Compared to the water at 1.03 specific gravity being pumped a small fraction of that height, should not be a limitation.
- The transport of the solids to the bottom of the trench could be done by basically a bucket elevator e.g., rail cars (at least 100 times the capacity of the present cars, i.e. a 5 fold scale increase in 3 dimensions or 125 fold) with return. Such delivery and return may preferably be encased in piping.
- As mentioned initially some form of **velocity enhancement** (more leverage per Archimedes) should be included, and by moving continental shelf mass down to the continental trench at say 45 miles/hour is a velocity enhancement, over the 2.5 cm/year movement of the continent, of 2*10^10. Combined with the head advantage of 50 fold yields an overall advantage of 10^12 over collecting the energy at a fault line (2, 11).
- Transfer of power from the rail car chain to the electric turbines could be by mechanical gearing or by suitable sealing around the rail cars and a larger diameter return pipe conduit (up to 2.5/1.0 times the area of the delivery pipe and its discharge and specified elevation above sea level) to supply electric turbines. One could simply use the volume of water returned in the emptied rail cars by extending the return pipe in elevation of say 1000 to 5000 feet above sea level. Other alternatives may include magnetic levitation of the buckets (rail cars) and one supposes magnetic braking could recover the electrical energy without need for a separate turbine/generator/power station.
- Mining of the sand deposits on shore may not be necessary. Hydraulic dredging of offshore deposits as the source (mine) and sand slurry transfer to the pipeline system for delivery to the continental trench may be preferred. Combined with magnetic levitation and magnetic braking for electricity generation (maglev) the system could be out of sight and out of mind for whatever benefits that may be for financial and aesthetic appeal. Considering that there are continental trenches around continents the choice of location worldwide is almost infinitely variable. While it is true that larger facilities have lower costs there may be merit in multiple facilities for the various multiple countries that will be involved.
- Population density in the area being mined has positive and negative benefits. Lots of people help construction and product use. Zero people help minimize resistance to the project. Considering the scale of the project zero people would seem preferred and the Western Sahara would be preferred on land or under-seas for hydraulic operation.

Conservation of the Environment: Unless the entire world's continental drift is totally constrained the creation of mountains, volcanoes, rain, mud, sand, rocks etc., will be conserved over time. Mass is neither being created nor destroyed, rather some of it is being returned to where it was at another time. Basically material on the continental shelf is being returned to the continental trenches. What people do with the product power will either help or harm the environment. The creation of the power provides the opportunity to mankind to decide the environments' and their own fate.

What Scale is needed? For the entire United States requirement in 2006 of 4.5*10^12 kwh or 600 gigawatts generating capacity the facility would be 300 times the scale of three gorges project in China or the U.S. Grand Cooley dam project using similar water flow and head. With a 50 fold head advantage the scale would be 300/50= 6 times that of either of these facilities. Assuming a water flow of 3 fold increase for that facility would require an overall 2 fold increase over either of these facilities. The total world electric demand would be satisfied by 10 such facilities.

Where are the facilities located? Considering that there are continental trenches at the boundary of the continents and over millions of years of erosion by wind and water, that have accumulated on continental shelves, they can be almost any where one desires. At some locations like Chile the continental trench is close by (e.g. 60 miles to 20,000 ft deep water) as well as having high elevation mountains near the coast. Others have broader continental shelves. It is not unreasonable to consider the exits (mouth) of major large rivers but the continental trenches may be less deep and farther away. This article is not a geological survey of the most economic locations; rather it is to observe that there are a very large number of candidates.

When will these facilities be built? Probably the time frame is 100 years in the future, say 2120. Plants are seldom built to provide greater than 20 years of visualizable needed capacity so something like 5 generation of concept and engineering evaluation followed by plant expansion will likely occur.

How much will electricity cost from a plant that is 300 greater capacity than the at present world's largest? If $0.10 will buy a kilowatt hour today then in 2012 dollars it will cost $0.10/(300)^(1/2)=$0.006/kwh.

Why will Continental Drift Power (CDP) be preferred? Depletion of oil and gas reserves are forecast to diminish significantly in 100 years, but it will be scale of operation and the low cost associated with the largest single plant output that drives Continental Drift Power (CDP) over the options of thousands of oil/gas wells and thousands of miles of distribution piping for its delivery (basically a 1 dimension activity). Green energy is basically a one year at a time event (in 2 dimensions), yet the proposed continental drift energy source has been accumulating for millions of years. In a way it has a 10^6 scale advantage over any earthbound green energy (and is scalable in 3 dimensions). Energy sources that have advantages of scale and accumulation over time include continental shelves, coal (strip mining is essentially 2 dimensional), oil, gas and nuclear energy materials, and are concentrated forms of energy. The competition between these accumulated sources will continue to play out on a cost basis.

How will Continental Drift Power (CDP) come about. I suppose it is not fair to misappropriate Winston Churchill and say "*It is not the end*, nor even the beginning of the end, but it is the end of the beginning" but the long road ahead would seem to make it appropriate. The need for energy from something other than fossil fuels is well known but the scale of presently discussed alternatives are so low as to make them expensive and ineffective in meeting the needs of the future. In the end it will be the scale, and accumulation of energy that will make continental drift energy prevail.

Is CDP conserved and sustainable? I am reminded of deriving differential equations in college wherein:*Input – Output – Reaction=Accumulation.* The input output difference is our desired power with zero chemical reaction required and the accumulation is over millions and millions of years. Over similar time frames it is conserving and fun to rewrite the equation: *Input – Output – Reaction –Accumulation = 0.* In other words it is conserving and sustainable for the human race for some time to come.

Is there Historical president or perspective? Tearing down mountains and dumping them in the sea is not a new concept but doing so to generate power for the world may be. There is a centuries old fable (7) in China where mountains are moved for the convenience of the human residents. Using hoes and baskets the family set out to dig up the mountain and dump it in the sea. Persisting for generation after generation ad infinitum they expected to accomplish their task; the moral of the story being many tasks appear unachievable but with determination anything can be achieved. *It has also been exemplified in Mao Tse Tung's telling of "Yu Gong Yi Shan," in a very famous speech.*

No philosophical or political motive for this article is intended though it is agreed that it will be the people of the world that will cause the mountains to be moved to generate the power. It is, after all, people that build the facilities in 3 dimensions over time and consume the product that is produced.

Rather it is to emphasize that the Scale of this proposal is markedly greater than the fables (hoes and baskets, vs. hydroelectric dams and volumetric flows greater than major rivers) and that the accumulated material (ground up mountain over millions of years vs. the original solid mountain) provides the energy basis for and from moving mountains.

Vertical Farming:

The use of electricity for light source for farming in a vertical as well as areal dimension (2 horizontal) is not new and has been discussed in the Economist (9). 'Vertical farming: does it really stack up". The conclusion there is correct, namely that the economics of doing so demand large scale low cost electricity. Given large scale low cost electricity farming can become 3 dimensional and therefore can be expanded at a greater rate than the growth rate of the population. In the very long term feeding the population can become a limitation that can be avoided by making farming a 3 dimensional activity.

World Population:

Presumably the world energy needs will grow with its population. Regardless of whose estimate one uses a significant increase beyond present will be required (10). Production of energy will preferably be proportional to the population, or greater, simply emphasizing the need for very large power generating facilities. Development of continental drift power (CDP) can presumably meet the demand within the next 100 years where population may increase 4 fold and continental drift power (CDP) can increase by 300 fold for each facility and 3000 fold for the world. Enough margin for growth and reduced costs too.

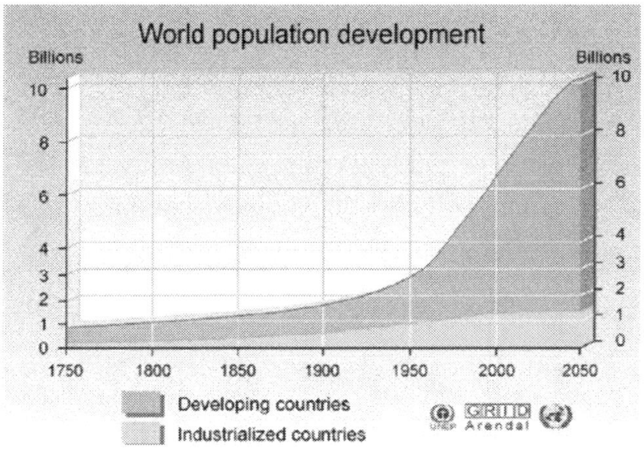

Will there be adequate Financial Capability: The financial capacity to do theses large projects will grow in time with the population hence should not be a limitation. For example the U. S. GDP alone may grow 10^4 over the 100 year period.

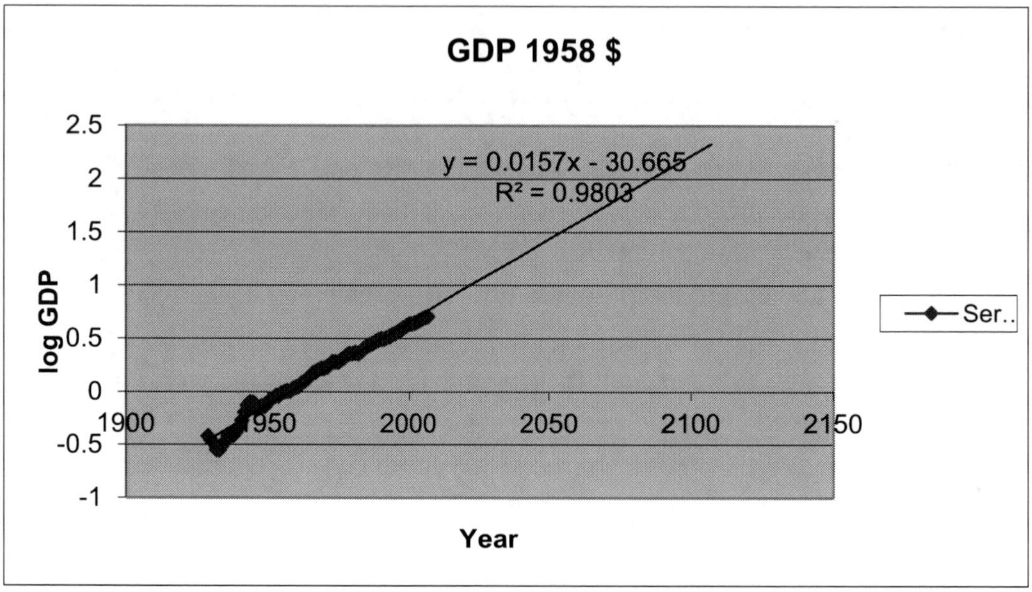

Other Benefits: All of the results (benefits) listed on page 2 naturally flow from very large scale availability of electric power at low cost. Nearly all are unattainable absent low cost power, e.g. the hydrogen economy, no green house gases, electric cars and airplanes, waste disposal reduced to total recycle, atmospheric modifications, vertical farming, even unlimited growth of the human race. And the visualizable list goes on limited only by ones imagination and can include space exploration, human colonization of planets and stars for the propagation of mankind etc. , etc.

Summary:

(1) The combined accumulated energy sources that make power from continental drift possible are:

- The radioactive decay of elements in the earth's interior that cause continental drift.
- The sun powered wind that erodes the mountains, evaporates the seas and causes rain to fall.
- The potential energy from elevated mass acted upon by gravity.
- Subduction Zones. When two oceanic plates collide, the younger of the two plates, because it is less dense, will ride over the edge of the older plate.
- Continental drift power (CDP) has tremendous scale, ergo cost, advantages vs. other alternative power sources.
- Continental drift power (CDP) is conservative and sustainable.

(2) A velocity enhancement over continental drift of 10^10 is achieved by gravitationally transporting mass to trenches at a higher velocity than the continental drift rate.

(3) The low cost of Continental Drift Power (CDP) owing to its scale will become the preferred alternative power source.

(4) "*It is not the end*, nor even the beginning of the end, but it is the end of the beginning", per Churchill.

References:

1. *http://scienceworld.wolfram.com/biography/Archimedes.html*
2. This article: Copyright: Donald I. Garnett, 14206 San Felipe Dr., Corpus Christi, Texas, Tuesday, 26 April 2011.
3. *http://experiencecurves.com/ecblog/*
4. *http://library.thinkquest.org/17457/platetectonics/2.php*
5. *http://www.stockton.edu/~hozikm/geol/Courses/The%20Earth/Content%20Web%20Pages/Bugielski/webpage.htm*
6. *http://www.eia.gov/emeu/iea/Notes%20for%20Table%206_2.html*
7. *http://english.cri.cn/4426/2007/01/15/167@185195.htm*
8. *http://www.story-lovers.com/listschinesestories.html*
9. *http://www.economist.com/node/17647627*
10. http://www.sustainablescale.org/areasofconcern/population/populationandscale/quickfacts.aspx
11. *http://physics-edu.org/energy_from_continental_drift.htm*

(145a) Prediction of EHS Properties Using T.E.S.T

Todd Martinand Douglas Young, National Risk Management Research Laboratory, US EPA, Cincinnati, OH

The US EPA has developed tools (such as the WAste Reduction Algorithm) which aid in the development of sustainable chemical process designs. In order to utilize these tools one must be able to characterize chemicals in terms of their EHS (environmental health and safety) properties. Since experimental values are not available for all industrially relevant chemicals, one needs to be able to estimate these properties using***in silico***based methods. T.E.S.T. (Toxicity Estimation Software Tool) was developed to allow users to easily estimate toxicity values using quantitative structure activity relationship (QSAR) models. The hierarchical clustering QSAR methodology was selected because it can develop models for large, diverse datasets. This methodology utilizes Ward's method to divide a training set into a series of structurally similar clusters. The structural similarity is defined in terms of about 800 2-D physicochemical descriptors (such as connectivity and E-state indices). A genetic algorithm based technique is used to generate statistically valid QSAR models for each cluster. The toxicity for a given query compound is estimated using the weighted average of several different cluster models (assuming that the compound is within the domain of applicability of the cluster model). T.E.S.T. can estimate toxicity values for a variety of toxicity endpoints including acute aquatic toxicity, acute mammalian toxicity, fish bioaccumulation factor, developmental toxicity, and mutagenicity. TEST can also estimate physical properties including normal boiling point, density, water solubility, and thermal conductivity. These properties are important for green solvent design. The predictive ability for each endpoint was validated using an external validation set.

(145b) Coupling Hoist Scheduling and Job Queue Optimization

Chuanyu Zhao, Jie Fu and Qiang Xu, Dan F. Smith Department of Chemical Engineering, Lamar University, Beaumont, TX

There are thousands of electroplating and polymeric coating shops in the U.S. annually producing numerous workpieces for many pillar industries. Hoist scheduling is commonly recognized one of the most important factors to improve their productivity. Nowadays, one challenging problem is that a plating shop should have the capability to simultaneously manufacture multiple types of jobs, and satisfy different custom requests. Common methods for this consideration include: i) a production line would be programmed to manufacture different types of job sequentially, i.e., processing the second type of job after fully completing the first type of job; ii) furnish multiple production lines and let one production line take care of a specific type of job. The first way is easy to set up but will cause significant delay for some job processing tasks. The second way can meet rapid manufacturing request but inevitably induce more investment and cost. To trade off between productivity and investment cost, a third way is proposed to mingle different types of job production in one single production line. Under the situation, the hoist is scheduled in a cyclic way to simultaneously handle multiple types of job so as to maximize the overall productivity. Conceivably, not only the hoist scheduling should be optimized, the best job loading queue to the production line should also be optimized.

In this paper, a systematic methodology coupling optimization of cyclic hoist scheduling and job loading queue for maximum productivity has been developed. It is programmed as an MILP (Mixed Integer Linear Programming) model addressing all the major concerned hoist scheduling issues, such as cyclic scheduling for large amount of jobs with multiple recipes, job queue determining, various production unit capacities, strict processing time requirements and exact manufacturing proportion for further assembling. This MILP model is programmed and solved in GAMS version 23.3 with the solver CPLEX. Since the model is fully linearized, the global optimal solution can also be guaranteed. The efficacy of the developed methodology is demonstrated by a case study for an electroplating system. The comparison for the cyclic hoist scheduling with and without job queue optimization is also conducted.

(145c) Material Handling and Production Line Debottleneck Optimization

Chuanyu Zhaoand Qiang Xu, Dan F. Smith Department of Chemical Engineering, Lamar University, Beaumont, TX

Multi-stage material handling (MSMH) processes are broadly used in industries for manufacturing massive amount of products/workpieces (jobs) through multiple processing steps, e.g., an electroplating system. A hoist is a controlled robot mainly conducting job lifting, releasing, and moving along its production line by following a preset movement schedule based on the job processing recipe. Nowadays, one of the big challenges for MSMH process is simultaneously manufacturing multiple types of jobs with different processing recipes from a single production line; meanwhile, some processing stages should have the multi-job processing capacities. These stringent requirements make optimal hoist scheduling become very complicated but also very important. In this paper, we consider such a situation that an MSMH production line has an optimal hoist schedule but the scheduling performance already meets an inherent bottleneck due to the lack of processing stages/units. Under this situation, additional parallel processing stages/units have to be added to boost the productivity associated with the development of a new hoist scheduling strategy, i.e., debottleneck the performance of the existing production line.

Unlike previous hoist scheduling cases, this material handling and production line debottleneck problem requires both decision making of the debottleneck scheme and hoist scheduling identification, which has never been discussed before. In this paper, an MILP (Mixed Integer Linear Programming) based modeling methodology is developed to optimize the debottleneck problem, including identifying which unit/units should be increased by their processing capacity and what is the optimal hoist schedule under the newly designed MSMH process, so that an investment will receive the best economic return. The efficacy of the proposed methodology is demonstrated by successfully tackling a cyclic hoist scheduling problem, where 8 units with various capacities are employed in the production line to continuously produce three different types of jobs. The scheduling result can be guaranteed as the global optimal solution as well.

(145d) Assessment of a Continuous Powder Mixing Process Using Online NIR Spectroscopy

Leonel Eduardo Quinones- Fontalvo, Chemical Engineering, University of Puerto Rico - Mayaguez Campus, Mayaguez, PR and Luis Guillermo Obregon-Quiñones, Chemical Engineering, University of Puerto Rico, Mayagüez Campus, Mayagüez, PR

NIR spectroscopic was used to get on-line spectra and obtain immediate results of API concentration at the output of a continuous powder mixer in order to assess the efficiency of the mixing. The results were compared with values obtained with ultraviolet spectrophotometer. Calibration curves were developed prior to continuous mixing experimentation for both UV and NIR methods. Considering that the mixing experimentation was performed between 0 and 30%w/w of Naproxen, it was created six different Partial least squares (PLS) calibration models for NIR ranging from 0 to 30% instead of using only one calibration model so that a better prediction were obtained. The NIR precision and exactitude achieved were about 2-5% and 1-3% w/w respectively depending on the value of concentration studied. The calibration models were generated taking into consideration the variability of the process working at different RPMs of the mixer and at different feed rates of the feeders and considering as well the powders properties mainly the powder cohesion variation in the mentioned range of API concentration. During the process of mixing three different step changes were applied to the feeders to obtain three different API concentrations in the range 0-30% to characterize the dynamic response of the continuous mixer. The results obtained by UV and NIR showed stable mixing at the different API concentrations and achieved a reasonable mixing with values of RSD lower than 6%. The error obtained with the NIR and UV increases when increasing the API concentration and it was attained to the increase in powder cohesion which made more difficult the process of mixing. The parameters of the dynamic response were similar to those obtained theoretically.

(145e) Method of Measurement of Carbon Permittivity

Manuel Guillermo Peña Benavides,**Dairo Andrés Vidal Teherán**and Santander Bolivar Solano Sr., Atl"ntico, Universidad del Atléntico, Barranquilla, Colombia

Coal is the most abundant fuel, versatile and available in the earth, global stocks are expected to continue until 250 years, so it becomes the most important energy source as it has been throughout its since history began to be implemented on locomotives and other transportation that worked with coal, promoted for the industrial revolution.

The use of coal as raw material requires a prior analysis of their composition that allows knowing the best process that should be used. Moreover, the complexity of coal in general, is such that there may be a wide variation of certain properties, even in coals that have the same origin or treatment, so it is essential to keep a detailed analysis of their composition and properties. Between these analysis are the dielectric properties, as is the permittivity, which is a physical constant that describes how an electric field affects and is affected by a medium, the permittivity is determined by the capacity of a material to polarize to the application of an electric field and thus partially cancel the internal field of the material, this property set to the behavior that will have the coals when exposed to an electromagnetic signal, marked as it spreads the energy in the material, as transforms the energy supplied by the microwave to heat the inside of it and, finally, indicates whether a material can be treated by means of microwave, providing valuable information for process heating and drying by microwave, which is the most viable way of warming and greater projection into the future because it provides even heating and fast, as it interacts with the material directly heats from the inside causing an agitation of the molecules which facilitates and speeds up the warming process.

Previous studies have identified several methods for the characterization of dielectric properties among which are the techniques of time domain reflectometry (TDR), the method of the guide shorted, perturbation method and some as the resonant cavity, S parameter method or the reflected wave method. These measurement methods have yielded data, processes and information that form the basis for further studies and designs, although still not enough for having performed for a frequency range or a specific frequency but difficult to achieve or is not handled.

The method of measuring the absorbed power, is proposed in this study, which is to apply to the test material a power known as a power which is provided to measure the amount of power absorbed by the material and then through a mathematical model to calculate the value of the permittivity.

In a first step to achieve this, it is necessary to use a device of waveguide tube has a rectangular cross section that is responsible for directing the microwaves from the generator to the sample. The device has the necessary instruments for measuring temperature and power consumption as well as a regulator of these two properties. This equipment is calibrated to identify what should be the power and heating time of the operation. This calibration is performed with distilled water and varying the power supplied to develop calibration curves to determine the most suitable operating power for the measurement of coal samples. These curves show the profile of heating and cooling. After an analysis of these is determined, in addition, the heating time, which is the time during which the coal samples are exposed to the action of microwaves.

It implements also an antenna inside the waveguide for measuring the intensity of the magnetic field generated by the device, because this value is of vital importance for the calculation of the permittivity.

With all this, the equipment can be standardized, which works with microwaves at a fixed frequency of 2.45 GHz.

The samples characterized in this study are of different types of coals are subjected to a pretreatment of grinding and sieving to obtain a homogeneous sample and is handled the same particle size for all types of coals analyzed. Also, for the analysis to be as accurate as possible all samples have the same mass and to the extent possible the same initial temperature.

The sample is introduced into an assay tube at the same time is introduced into the device and is subjected to a heating power and time. Parallel to the main sample is being heated by microwaves, another sample of equal mass and size of particle is heated by a resistance system that is placed within the respective assay tube and connected to a source of energy electrical and within this circuit are connected to each amperage and voltage meters and heated in the same proportion of the sample temperature main analysis, so we can determine the amount of power absorbed by the sample parallel would equal the main sample.

The temperature of both samples is measured by an infrared thermometer strategically located to capture directly the heat of the respective sample.

After obtaining the data of power consumption, temperature, power supplied, magnetic field strength and some physical properties of the material under study, can proceed to calculate the permittivity through a mathematical model to relate these data and properties. This procedure is repeated several times to compare with each other so that they can be correlated and get a more accurate value of permittivity.

Keep in mind that these values of the permittivity have a frequency of 2.45 GHz for very few substances. The scope of this paper is precisely that, to be able to provide data that are not known and knowing that at that frequency is working commercial microwave generators and this is the best way of warming, these data become a even greater importance and can help large industries to reduce costs, reduce processing times, have a more controlled process or implement new technologies in them.

Another advantage of this method of measuring the permittivity is that it is carried out at low cost, while the existing and currently being used have a very high cost of operation or equipment used in these processes are very expensive.

In this paper we also hope that later you can implement this process for calculating the permittivity of solid substances, and why not? Adjust the equipment to find the property to liquid substances, data are not yet known.

(145f) Optimization Tools for Development of Unconventional Natural Gas Resources

Srimoyee Bhattacharya, University of Houston, Houston, TX and M. Nikolaou, Chemical and Biomolecular Engineering, University of Houston, Houston, TX

Massive hydraulic fracturing of horizontal wells has made the recovery of natural gas from unconventional gas reservoirs both technically and economically feasible. However, the design of hydraulic fracturing relies on extensive what-if analysis using commercial simulators. The entire design process is fairly demanding, and may lead to designs that are far from optimal. In this work, we present a computer-aided approach that facilitates engineers to efficiently converge to solutions that are at or close to the optimum.

The proposed optimization objective is the net present value (NPV) for horizontal well drilled in the direction of the minimum horizontal stress, which allows multiple transverse fractures. Primary decision variables are the number of transverse fractures in a horizontal well; horizontal well length; fracturing fluid viscosity; proppant concentration; injection rate of fracturing fluid; and injection time. Key constraints are fracture growth control, operational constraints, and geometric constraints. Critical reservoir properties (e.g. permeability, porosity, and stress orientation) needed for optimization is assumed to be available from a variety of sources (e.g., seismic, prior production logs). The PKN model[1] for hydraulic fracturing propagation is used which is suitable for long fractures. The productivity of multifractured horizontal wells is computed by proposing a new analytical method and compared with a numerical simulator for consistency. The optimization algorithm used is GLOBAL developed by Csendes[2] which is a multistart clustering global optimization method . Probabilistic optimization algorithms are also used for comparison.

The proposed approach is used to design horizontal well geometry and stimulations for a number of test cases, based on numerical simulations. Results are compared with other available case studies in literature. The proposed approach demonstrates significant increase in NPV over other available methods. The simulations indicate good numerical efficiency and ability of the approach to be used as a decision making tool for unconventional gas field development. Work is currently underway to integrate this tool with vertical well hydraulic fracture optimization.

References:

1. Perkins, T.K., and Kern, L.R, 1961. Widths of Hydraulic Fractures. Journal of Petroleum Technology: 937-49.

2. Csendes, T., Pal, L., Sendin, J.H., Banga, J.R., 2008. The GLOBAL optimization method revisited. Optimization Letters, 2(4): 445-454.

(149b) Plant Equipment Overdesign

Mohamed Soliman, NATSD, PE, AICHE, Ras Tanura, Saudi Arabia

Many Engineers overdesign plant equipment, valves and pipes to account for uncertainty even though they have designed the same equipment a hundred times. An acceptable and encouraging name is given to the overdesign factor for example, safety margin or design factor. It starts with the proponent (operating company), through the design contractor and finally the vendor. It is the mentality of, "the bigger the better." If it is oversized, it will work. However, if it is undersized, it will not work and everyone will hear about it and it will be a lesson learned to everyone.

One of the main reasons for overdesign factor is attributed to the lack of clear guidelines and instruction from the proponent to take responsibility of not adding this factor.

The design factor increases the operating costs and consequently reduces the plant profitability. It also affects the plant safety. Liquid hammering in the plant could be attributed to pipes overdesign.

In this paper we will discuss the reasons for adding an overdesign factor and the consequences of overdesigning the plant equipment, pipes and valves with real life case studies. Oversizing the crude charge pump in one of the Gas Oil Separating Plant (GOSP) will be discussed in details. The oversizing did not only waste energy, but it also interrupted the plant operation and caused the continuous opening of off-spec valves.

Scaling Batch and Semi-batch Processes from Laboratory to Plant

Jonathan H. Worstell

Abstract

The heart of any chemical process is the reactor. The most common reactor types are batch and semi-batch. The efficiency of both reactor types depends upon agitation; i.e., mixing. Thus, successfully scaling batch and semi-batch reactors depends upon mixing. A plethora of mixing correlations exist; unfortunately, hardly any two batch or semi-batch reactors are geometrically similar enough for such correlations to work well.

This paper advocates using mass transfer coefficients as a means for scaling batch and semi-batch reactors from laboratory to plant. The majority of this paper discusses semi-batch reactors.

Introduction

The reactor is the "heart" of every chemical process. Why --- because molecular change occurs in the reactor. Physical change occurs in all other equipment in a chemical process; impurities are absorbed or adsorbed from feed streams, thereby purifying the reactant; product is evaporated, then condensed, thereby separating it from a solvent or by-product-contaminated stream; solvent or by-product or both are volatilized, then condensed, thereby separating them from the product stream; extraction can also be used to separate product from solvent and by-products. In these examples, change occurs physically, not chemically: molecules do not change composition; only their physical state changes. Thus, to be a chemical process, the process must contain a chemical reactor.

Tank reactors are easy to build; therefore, they are relatively inexpensive. They have a high utilization rate since they are flexible: they can be used as mixers, separators, and storage vessels, as well as chemical reactors. All tank reactors possess agitators. Agitators can be mounted atop, under, or, or on the side of a tank reactor. Tank reactors charged once, then discharged after product formation, are batch reactors. Tank reactors that are charged with a liquid phase containing a reactant, then continuously charged with a second reactant to control reaction temperature are semi-batch reactors. Tank reactors that are continuously charged and discharged are continuous stirred tank reactors (CSTR's).

In semi-batch processes, the chemical component with the higher heat capacity is charged to the tank reactor while the second chemical component is pumped into the reactor at the same rate that heat is removed from the reactor. The reactant pumped into the reactor may be a gas, an immiscible liquid, or a miscible liquid. In the first case, gas bubbles populate the liquid phase present in the reactor. In the second case, liquid droplets; i.e., micelles, populate the liquid phase present in the reactor. In the third case, the reactor contains only one fluid, a liquid containing reactants and products.
Scaling tank reactors is generally a matter of heat transfer, so long as the reaction rate is the only concern. If a precipitate forms during the reaction, then particle size or particle shape may be important. Scaling precipitate characteristics is difficult. Scaling semi-batch reactors is also difficult because of the various phases in the reactor. Using dimensionless parameters related to the geometry of the reactor and using the power number or power/volume ratio constitute the most widely accepted methods for scaling such reactors. This paper discusses the physics and scaling parameters primarily for semi-batch reactors and proposes an empirical method for scaling them.

Semi-batch Reactors

Figure 1 depicts the basic features and design of a gas-liquid semi-batch reactor. A baffled tank is fitted with a turbine-style agitator that evenly distributes gas bubbles throughout the liquid phase contained by the vessel. The gas enters the tank through a pipe attached to a "frit" projecting under and parallel to the turbine in the tank. The same design can be used for mixing and reacting immiscible liquids. For miscible reacting liquids, the second liquid will be sprayed onto the agitated surface of the contained liquid. If a miscible liquid is injected into the mass of the agitated liquid through a frit, the frit will more often than not plug as product forms upon immediate contact of the two liquids. Frit plugging will certainly occur if the product is solid or highly viscous. If the process requires that the miscible liquid be injected into the mass of the agitated liquid, the openings on the injection nozzle must be large enough to ensure that they do not plug with product solid or become congested with viscous product.

The gas or immiscible liquid will form a coherent mass in the agitated liquid. Inside this coherent mass, reactant molecules diffuse in all directions amongst themselves. However, at some distance from the gas-liquid interface, the direction of reactant diffusion becomes oriented toward the interface. This region of oriented diffusion is called the "stagnant film" of the gas bubble or of the micelle. Diffusion is oriented in this stagnant film because reactant is moving from a region of high concentration; i.e., the interior of the gas bubble or liquid micelle, to a region of low reactant concentration --- that is: to the physical interface between the gas or immiscible liquid and the agitated liquid. No convection occurs inside the gas bubble or liquid micelle; therefore, the breadth of the stagnant film only depends on the volume of the bubble or micelle.

A similar stagnant film exists on the agitated liquid side of the bubble or micelle. This stagnant film exists due to the stipulation that the velocity of a flowing fluid touching a surface is zero. This stipulation is basic to fluid dynamics: airplanes fly because of it. In this liquid stagnant film, reactant molecules flow from the gas-liquid or liquid-liquid interface in an oriented fashion toward the bulk liquid, where they react to form product. Product forms in this stagnant film if the reaction is fast. The agitation rate; i.e., bulk fluid velocity, determines the breadth of the liquid stagnant film: high fluid velocity produces a thin stagnant film while low fluid velocity produces a thick stagnant film.

Physics of Gas-liquid Semi-batch Reactors

The movement of gas molecules across the gas stagnant film occurs via diffusion. Thus, it can be modeled as a first-order rate equation.

$$R_{GF}=k_G(P_{GasBubble} - P_{Interface})$$

where R_{GF} is the rate of gas molecule movement (moles/m$^3\cdot$hr); k_G is the rate constant for such diffusion (moles/m$^3\cdot$hr$\cdot$atm); $P_{GasBubble}$ is the pressure of the gas bubble (atm); and, $P_{Interface}$ is the gas pressure at the gas-liquid interface (atm). Henry's Law controls the amount of gas dissolving into the liquid at the gas-liquid interface. Henry's Law is

$$He C_{Eq} = P_{Interface}$$

where He, Henry's Law constant, has dimensions of $m^{3}\cdot atm/mole$ and C_{Eq} is the equilibrium liquid phase concentration of reactant at the gas-liquid interface. C_{Eq} has units of $moles/m^{3}$. Dissolved reactant molecules move across the liquid stagnant film via diffusion; therefore dissolved reactant can be modeled as a first-order equation.

$$R_{LF}=k_{L}a_{L}\left(C_{Eq}-C_{Bulk}\right)$$

where R_{LF} is the rate of dissolved reactant movement ($mole/m^{3}\cdot hr$); k_L is the rate constant diffusion in liquid (m/hr); a_L is the surface area to volume ratio of the liquid stagnant film (m^2/m^3); $C_{Interface}$ is the reactant liquid concentration at the gas-liquid interface ($mole/m^3$); and, C_{Bulk} is the concentration of dissolved reactant where the liquid stagnant film becomes indistinguishable from the bulk fluid; i.e., where the fluid velocities for adjacent fluid layers are equal.

There are only two measured quantities in the above equations: P_G and C_B. For a slowly reacting system, the concentration gradient within the liquid stagnant film will be well developed. Solving the above rate equations for P_G and C_B yields

$$\frac{R_{GF}}{k_G He}+\frac{R_{LF}}{k_L a_L}=\frac{P_G}{He}-C_{Bulk}$$

The observed overall rate equation is

$$R_{Overall}=k_{Overall}\left(\frac{P_G}{He}-C_{Bulk}\right)$$

Rearranging yields

$$\frac{R_{Overall}}{k_{Overall}}=\left(\frac{P_G}{He}-C_{Bulk}\right)$$

Substituting the above result into the third equation above gives

$$\frac{R_{GF}}{k_G He} + \frac{R_{LF}}{k_L a_L} = \frac{R_{overall}}{k_{Overall}}$$

Assuming $R_{GF} = R_{LF} = R_{Overall}$, then the above equation becomes

$$\frac{R_{Overall}}{k_G He} + \frac{R_{Overall}}{k_L a_L} = \frac{R_{Overall}}{k_{Overall}}$$

Thus

$$\frac{1}{k_G He} + \frac{1}{k_L a_L} = \frac{1}{k_{Overall}}$$

The above equation is in the form of a "series resistance" mechanism. It shows that the overall; i.e., total, resistance to moving gaseous reactant to dissolved reactant is the sum of resistance to gas molecules crossing the gas stagnant film and the resistance to dissolved reactant molecules crossing the liquid stagnant film.

So --- which resistance is the most important, the major contributor to the overall resistance? The Sherwood number is defined as

$$N_{Sh} = \frac{\text{Total Mass Transfer}}{\text{Molecular Mass Transfer}} = \frac{k_{g\text{ or }L}\delta}{D}$$

where $k_{G\text{ or }L}$ is the rate constant for gas or liquid molecules crossing a stagnant film (m/hr); δ is the breadth of the stagnant film (m); and, D is diffusivity (m^2/hr). For spheres, the Sherwood numbers for gas molecules and for liquid molecules possess similar orders of magnitude. Thus

$$O\left(\frac{k_G L}{D_G}\right) \cong O\left(\frac{k_L L}{D_L}\right)$$

The diffusivity for a gas is much higher than the diffusivity for a liquid because the mean free path of a gas is longer than that of a liquid. Therefore, for the above order of magnitude relationship to hold

$$k_G \gg k_L$$

Thus, the controlling resistance is $^1/_{k_L a_L}$.

Evaluation of $k_L a_L$

To reduce the overall resistance, $1/k_{Overall}$, requires a reduction of the resistance $1/k_L a_L$. In other words, $k_L a_L$ must increase to increase the conversion rate for the process. Chemical engineers evaluate $k_L a_L$ as a unit. The most common methods for evaluating correlations, one such being

$$k_L a_L = K\left(^{Pr_G}/_V\right)^{\alpha}(v_S)^{\beta}$$

where Pr_G/V is a gassed power to unit volume ratio; v_S is superficial gas velocity; and, K is a constant with the appropriate dimensions to balance the equation.[1] α and β are determined by the specific correlation. One study gives them as $\alpha = 0.475$ and $\beta = 0.4$.[2] Another study found $\alpha = 0.4$ and $\beta = 0.5$ for water volumes to 2600 liters.[3] Yet another study reported $\alpha = 0.54$ and $\beta = 0.68$.[4] The reported differences for α and β may not seem great; however, they are exponents, thus they can produce a calculated result containing significant error. Also, perusing the literature shows that α and β depend on

---

[1] J. M. Smith, K. van't Riet, and J. Middleton, *Proceedings of the Second European Conference on Mixing*, Cambridge, England, BHRA Fluid Engineering, Cranfield, England, F4-51, 1977.
[2] *Ibid.*
[3] K. van't Riet, *Industrial and Engineering Chemistry: Process Design Development*, *18*, 357 (1979).
[4] A. D. Hickman, Proceedings *of the Sixth European Conference on Mixing*, Pavia, Italy, BHRA Fluid Engineering, Cranfield, England, 369, 1988.

the bulk fluid and its properties, as well as vessel geometry and size. For example, an acetic acid fermentation medium had $\alpha = 0.48$ and $\beta = 0.74$.[5]

a second correlation is embedded in the above k_La_L correlation. That second correlation is for gassed power, Pr_G. The standard correlation for gassed power is

$$Pr_G = c\left(\frac{Pr_0^2 Q^{0.44}}{N_A}\right)^{0.45} = c\left(\frac{Pr_0^2 ND^3}{Q^{0.56}}\right)^{0.45}$$

where, in the first bracket, Pr_0 is ungassed power, Q is gas sparging rate, N_A is aeration number

$$N_A = \frac{Q}{ND^3}$$

where N is impeller rotational speed, D is impeller diameter; for both brackets, c is a constant dependent on impeller geometry.[6] The below correlation determines Pr_0

$$Pr_0 = \frac{\rho N_P N^3 D^5}{g_c}$$

where ρ is density, N_P is ungassed power number, and g_c is the gravitational constant. N_P is given by

$$\frac{Pr_0 g_c}{\rho N^3 D^5} = N_P$$

The last two equations above simply imply that Pr_0 has to be measured. Other correlations have been proposed for Pr_G, such as[7]

---

[5]J. Baczkiewica and H. Michalski, *Proceedings of the Sixth European Conference on Mixing*, Pavia, Italy, BHRA Fluid Engineering, Cranfield, England, 473, 1988.
[6] B. Michel and S. Miller, *AIChEJ*, *8*, 262 (1962).
[7] S. Nagata, *Mixing: Principles and Applications*, Halsted Press, New York, NY, page 60, 1975.

$$\ln\left(\frac{Pr_G}{Pr_0}\right)=-192\left(\frac{D}{T}\right)^{4.38}\left(\frac{\rho D^2 N}{\mu}\right)^{0.115}\left(\frac{N^2 D}{g}\right)^{1.96(D/T)}\left(\frac{Q}{ND^3}\right)$$

where T is tank diameter, μ is liquid viscosity, and g is gravitational acceleration. Another correlation has been proposed[8]

$$\frac{Pr_G}{Pr_0}=CN_{We}^{x}N_A^{y}\left(\frac{\rho_L}{\rho_D}\right)$$

where C is a constant, ρ_L is continuous liquid phase density, ρ_D is dispersion density, N_{We} is the Weber Number, given as

$$N_{We}=\frac{N^2 D^3 \rho_L}{\sigma}$$

σ is interfacial surface tension. Yet another gassed power correlation is[9]

$$Pr_G=0.83\left(\frac{Pr_0^2 ND^3}{Q^{0.56}}\right)^{0.45}$$

Power is energy per time and in a chemical reactor, ungassed power is the energy per unit time imparted to the liquid fluid by the mechanical agitator. Gassed power is the energy per unit time entering the liquid via the mechanical agitator and the sparging gas. Determining gassed power is much more difficult than determining ungassed power.

Much depends on power --- blend time, phase holdup, dispersion bubble or micelle size and bubble or micelle distribution, reaction times for fast chemical reactions, mass transfer coefficients, heat transfer coefficients, particle size and shape

[8] I. Hassan and C. Robinson, *AIChEJ*, *23*, 48 (1977).
[9] B. Loiseau, N. Midoux, and J. Charpentier, *AIChEJ*, *23*, 894 (1977).

for precipitate-forming chemical reactions, sustainability of biochemical reactions in microorganism suspensions, and utility costs. Thus, much depends on the quality of power correlations since they are used for design purposes and scaling purposes in the CPI. However, power depends significantly on reactor geometry, not just on tank diameter and height, or impeller diameter, or impeller clearance with regard to tank bottom and tank wall, but also on all the coils and nozzles and dip tubes and baffles projecting into the liquid present in the tank --- which means that most mixing correlations are valid only for the tank used for data collection. Thus, it is no surprise that differences between power need as calculated using a correlation and actual power need occur. Another reason for differences between calculated power need and actual power need is agitator operating efficiency. The integral mechanical features of an agitator; i.e., seals and bearings, need, consume energy, but such energy need is not represented in most power correlations, such as the ones above. For a well-mixed batch reactor, the power efficiency may only be sixty or eighty percent.[10] For these reasons, criticism of k_La_L-Pr correlations has developed.[11] This criticism suggests that the k_La_L-Pr correlation should take the functional form

$$k_La_L=f\left(\frac{Pr_T}{V},\ v_S,\ \text{fluid parameters, coalescence parameters}\right)$$

where Pr_T total power input to the reactor.

The most disconcerting feature of these types of k_La_L correlations is that they employ physical parameters not related to k_La_L. First, k_L is the rate at which molecules move through a given fluid layer. Therefore, k_L is related to the mean free path of molecular translation divided by time. Fluid density impacts mean free path: in general, mean free path decreases with increasing density. When density does appear in k_La_L-Pr correlations, it is as a “weak” independent variable; i.e., it is taken to a low exponential power. Second, a_L is the ratio of liquid stagnant film surface area to liquid stagnant film volume. The physical parameter most closely related to a_L is bubble

---

[10] G. Tatterson, *Scaleup and Design of Industrial Mixing Processes*, McGraw-Hill, Inc, New York, NY, Page 54, 1994.
[11] V. Linek, P. Benes, and O. Holecek, *Biotechnology and Bioengineering*, *32,* 482 (1988).

diameter or micelle diameter, neither of which appears universally in k_La_L-Pr correlations.

Scattered throughout the literature pertaining to k_La_L-Pr correlations are comments implying k_L is independent or, at best, weakly dependent upon such correlations. Mass transfer rate constants can be defined as

$$k_L = {D_{Diff}}/{\delta}$$

where D_{Diff} is molecular diffusivity (m^2/hr) and δ is stagnant film breadth (m). Thus, k_G and k_L depend linearly on D_{Diff} and inversely on δ. D_{Diff} depends upon process temperature; therefore, it is constant for a given process. For the liquid phase, δ depends upon the velocity of the bulk liquid streaming past the bubble or micelle; therefore, it depends upon the Reynolds number, N_{Re}, for the process: as N_{Re} increases, δ decreases, thereby increasing k_L. Some k_La_L-Pr correlation relate the impeller Reynolds number to k_La_L, as in this equation from above

$$\ln\left(\frac{Pr_G}{Pr_0}\right) = -192\left(\frac{D}{T}\right)^{4.38}\left(\frac{\rho D^2 N}{\mu}\right)^{0.115}\left(\frac{N^2 D}{g}\right)^{1.96\left({D}/{T}\right)}\left(\frac{Q}{ND^3}\right)$$

where ${\rho D^2 N}/{\mu}$ is the impeller Reynolds number. Note, however, that impeller N_{Re} is not necessarily bulk fluid N_{Re}, although as impeller N_{Re} increases, so must bulk fluid N_{Re}. Note also, that impeller N_{Re} is only weakly correlated to k_La_L, as indicated by the low exponential power.[12]

Also, comments appear suggesting that a_L increases with increasing impeller RPM. Such comments are valid: increasing the velocity of the bulk fluid streaming past the bubble or micelle reduces the breadth of the liquid stagnant film, thereby decreasing

[12] The weak dependence between k_La_L and impeller N_{Re} in this correlation implies the process studied was not diffusion rate controlled, but rather reaction rate controlled. See below.

the volume of the stagnant layer while not appreciably impacting the surface area of the layer. Thus, a_L increases with increasing bulk fluid velocity, thereby increasing $k_{Overall}$.

Now reconsider the equation for a liquid stagnant film controlled process: it is

$$\frac{1}{k_G He}+\frac{1}{k_L a_L}=\frac{1}{k_{Overall}}$$

where k_G and k_L are constants depending upon mean free path translation divided by time and He is Henry's Law constant, which temperature and pressure specify. Thus, $k_{Overall}$ only depends upon a_L. Since chemical processes operate at constant temperature and constant N_{Re}, the above equation represents a straight line where $1/k_G He$ is the intercept and $1/k_L$ is the slope, as shown in Figure 4.

Investigating Semi-batch Chemical Processes

Chemical reactors with heterogeneous phases; i.e., with an interface between two distinct phases, possess three zones of operation: diffusion rate controlled, transition rate controlled, and reaction rate controlled. When diffusion rate controlled, the rate constant for reactant to product conversion depends upon the rate at which reactant diffuses across a stagnant film. In this control regime, potential reaction sites are starved for reactant. When reaction rate controlled, the rate constant for reactant to product conversion is the kinetic, reaction rate constant. In this control regime, all potential reaction sites are saturated with reactant. The transition control regime occurs when the diffusion rate nearly matches the reaction rate --- at one moment the process is diffusion rate limited, at the next moment, it is reaction rate limited. The most efficient and productive control regime for a chemical reactor is the reaction rate regime. That is the operating target when scaling a chemical reactor.

The above equation and Figure 4 show that a_L is the most important dependent variable when optimizing the operation of a gas-liquid or immiscible liquid-liquid semi-batch reactor. Measure $k_{Overall}$ at constant temperature, pressure, and RPM for calibrated frits, then chose the optimum frit for further experimentation., Next measure $k_{Overall}$ as a function of impeller RPM at constant temperature, pressure, and bubble or micelle diameter; i.e., using the optimum frit determined in the first experimental set. Figure 5 depicts the data from such an experimental set. When $k_{Overall}$ depends on impeller RPM, then the reactor is operating in the diffusion rate control regime. When $k_{Overall}$ is independent of impeller RPM, then the reactor is operating in the reaction rate control regime, which means all potential reaction sites are functioning at maximum efficiency and productivity. The region between diffusion rate control and reaction rate control regimes is the transition regime: avoid operating in it.

Chemistry of Gas-liquid Semi-batch Reactors

The $k_{Overall}$ measured for the diffusion rate controlled portion of Figure 5 is dominated by the physical movement of reactant across the liquid stagnant film. However, the liquid stagnant film is negligible in the reaction rate controlled portion of Figure 5. In the reaction rate controlled portion of Figure 5, the rate of gaseous reactant across the gas-liquid interface is the same as above, namely

$$R_{GF}=k_G(P_{GasBubble} - P_{Interface})$$

and Henry's Law still controls the amount of gas dissolving into the liquid at the gas-liquid interface: thus

$$HeC_{Eq} = P_{Interface}$$

In the reaction rate controlled regime, dissolved reactant does not move through a liquid stagnant film before reacting; chemical conversion can occur at the gas-liquid interface. Thus

$$R_R= k_R(C_{Eq}- C_B)$$

Where R_R is reaction rate (moles/$m^3 \cdot$hr); k_R is the reaction rate constant (1/hr), assuming a first-order chemical reaction; C_{Eq} is dissolved reactant concentration at the gas-liquid interface (moles/m^3); and, C_B is dissolved reactant concentration in the bulk liquid (moles/m^3). Solving the above chemical process in terms of measurable quantities; I.e., in terms of $P_{GasBubble}$, He, and C_B, yields

$$\frac{R_{GF}}{k_G He} + \frac{R_R}{k_R} = \frac{P_G}{He} - C_{Bulk}$$

The observed overall rate equation is

$$R_{Overall} = k_{Overall}\left(\frac{P_G}{He} - C_{Bulk}\right)$$

Rearranging yields

$$\frac{R_{Overall}}{k_{Overall}} = \left(\frac{P_G}{He} - C_{Bulk}\right)$$

Substituting the above result into the third equation above gives

$$\frac{R_{GF}}{k_G He} + \frac{R_R}{k_R} = \frac{R_{overall}}{k_{Overall}}$$

Assuming $R_{GF} = R_R = R_{Overall}$, then the above equation becomes

$$\frac{R_{Overall}}{k_G He} + \frac{R_{Overall}}{k_R} = \frac{R_{Overall}}{k_{Overall}}$$

Thus

$$\frac{1}{k_G He} + \frac{1}{k_R} = \frac{1}{k_{Overall}}$$

Thus, the reaction rate controlled regime can be represented as a series of resistances: resistance to solvating a reactant gas molecule and resistance to converting that dissolve reactant molecule to product molecule. Note that in this process regime, the only way to increase $k_{Overall}$ is to increase process pressure, thereby increasing $P_{GasBubble}$ and C_{Eq} via Henry's Law.

Chemistry and Physics of Transition Regime

The transition regime of Figure 5 is a combination of the resistances for diffusion rate controlled and reaction rate controlled regimes. In other words

$$\frac{1}{k_{Overall}^{Transition}} = \frac{\alpha}{k_{Overall}^{Diffusion}} + \frac{\beta}{k_{Overall}^{Reaction}}$$

where $\alpha+\beta = 1$. Substituting into the above equation yields

$$\frac{1}{k_{Overall}^{Transition}} = \alpha\left(\frac{1}{k_G He} + \frac{1}{k_L a_L}\right) + \beta\left(\frac{1}{k_G He} + \frac{1}{k_R}\right)$$

$$\frac{1}{k_{Overall}^{Transition}} = \left(\frac{\alpha}{k_G He} + \frac{\alpha}{k_L a_L}\right) + \left(\frac{\beta}{k_G He} + \frac{\beta}{k_R}\right)$$

Combining terms gives

$$\frac{1}{k_{Overall}^{Transition}} = \frac{\alpha+\beta}{k_G He} + \frac{\alpha}{k_L a_L} + \frac{\beta}{k_R}$$

or

$$\frac{1}{k_{Overall}^{Transition}} = \frac{1}{k_G He} + \frac{\alpha}{k_L a_L} + \frac{\beta}{k_R}$$

Thus, the process can shift from one control regime to another as the magnitudes of α and β change with small changes in operating conditions. Also, for a given set of operating conditions, portions of the bulk liquid in the reactor can be diffusion rate controlled while other portions of the bulk liquid in the reactor can be reaction rate controlled. For these reasons, it is best to avoid operating a semi-batch reactor in the transition regime.

Scaling Semi-batch Reactors

When scaling a semi-batch reactor, the first consideration is to maximize the gas bubble surface area to volume ratio or the liquid micelle surface area to volume ratio because all mass transport within the bubble or micelle is by diffusion; no convection currents exist within the bubble or micelle. The larger the bubble or micelle surface area to volume ratio, the higher reactant diffusion becomes. Sintered frit injectors are much more efficient at creating small bubbles or micelles than are agitators. Therefore, specify the smallest possible sintered frit openings for the process and install the sintered frit injector under the bottom turbine or propeller.

Second, the concentration of dissolved gas depends on Henry's Law. Therefore, operate the process at the highest possible pressure that is equipment safe. Not only does high gas pressure insure the highest possible dissolved reactant concentration, C_{Eq}, but it also induces jetting of gas from the frit nozzle, thereby increasing gas bubble dispersion and imparting more energy to the bulk liquid. The solubility of immiscible liquid reactant in bulk liquid is process temperature dependent; thus, operate the process reactor at the optimum temperature for obtaining maximum reactant solvation.

Third, bulk liquid flow past the bubble or micelle controls the breadth of the liquid stagnant film surrounding the bubble or micelle, thereby increasing the amount of dissolved reactant entering the bulk liquid. Therefore, operate the agitator in the turbulent zone of the process, as shown in Figure 6.[13] The turbulent zone for tank reactors is at impeller Reynolds Number, N_{Re}, greater than 10^4. Avoid the flow transition region of Figure 6; that is, avoid operating the process between 10^1 and 10^4 impeller N_{Re}. Operating in the flow transition region of Figure 6 makes it difficult to achieve consistent chemical reaction rates.[14]

---

[13] F. Holland and F. Chapman, *Liquid Mixing and Processing in Stirred Tanks*, Reinhold Publishing Corporation, New York, NY, page 78, 1966.
[14] G. Tatterson, *Scaleup and Design of Industrial Mixing Processes*, McGraw-Hill, Inc, New York, NY, Page 14, 1994.

To obtain a good dispersion, install four symmetrically spaced baffles in the semi-batch reactor. Doing so increases the power number for the process, as shown in Figure 7.[15] Note that wide baffles are better than narrow baffles. Also, use six flat bladed disk turbines to obtain good dispersion and impart more power to the bulk liquid, as shown in Figure 8.[16]

Finally, determine $k_{Overall}$ as a function of impeller RPM. If $k_{Overall}$ depends on impeller RPM, then a significant liquid stagnant film exists in the process and is adversely impacting process efficiency and productivity. Installing multiple turbines along the agitator shaft may be required to impart more energy to the flowing bulk liquid.

---

[15] F. Holland and F. Chapman, *Liquid Mixing and Processing in Stirred Tanks*, Reinhold Publishing Corporation, New York, NY, page 79, 1966

[16] R. Bates, P. Fondy, and R. Corpstein, *Industrial and Engineering Chemistry, Process Design and Development*, 2, 310 (1963).

Figure 1.
Depiction of a Gas-liquid Semi-batch Reactor.

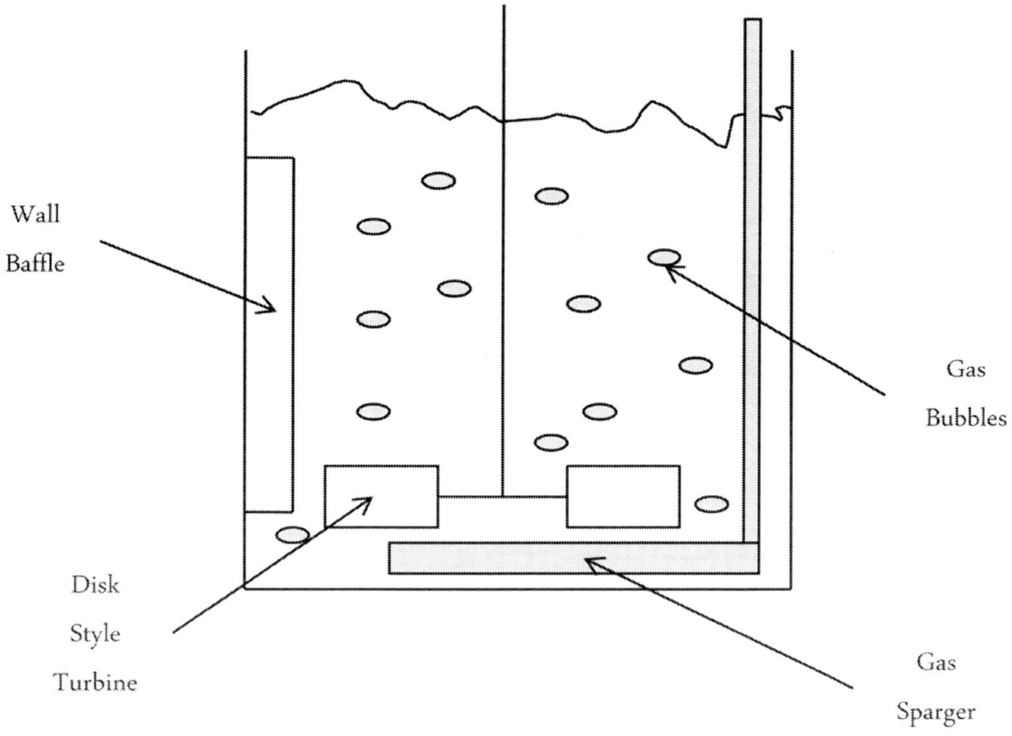

Figure 2.
Depiction of Gas or Immiscible Micelle in a Semi-batch Reactor.

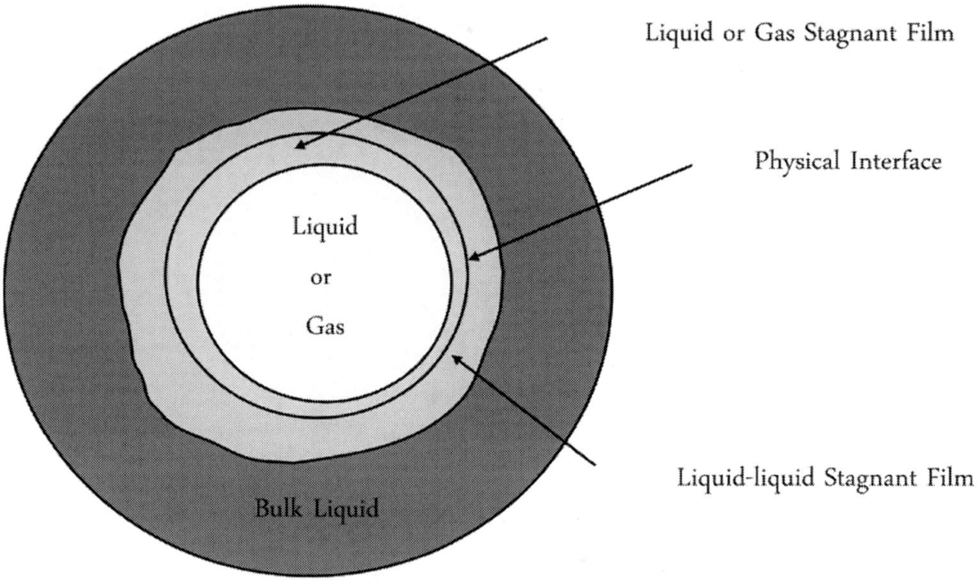

Figure 3.
Rate Equations for a Gas Bubble or Immiscible Micelle in a Semi-batch Reactor.

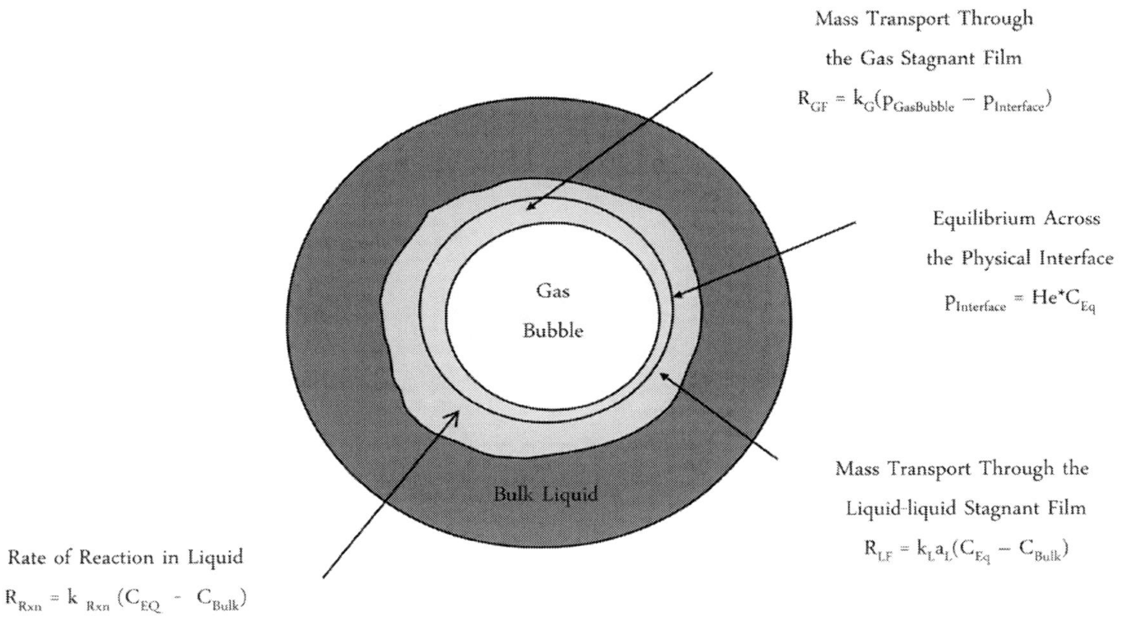

Figure 4.
Relationship between $1/k_{Overall}$ and $1/a_L$ for a
Liquid Stagnant Film Controlled Chemical Process.

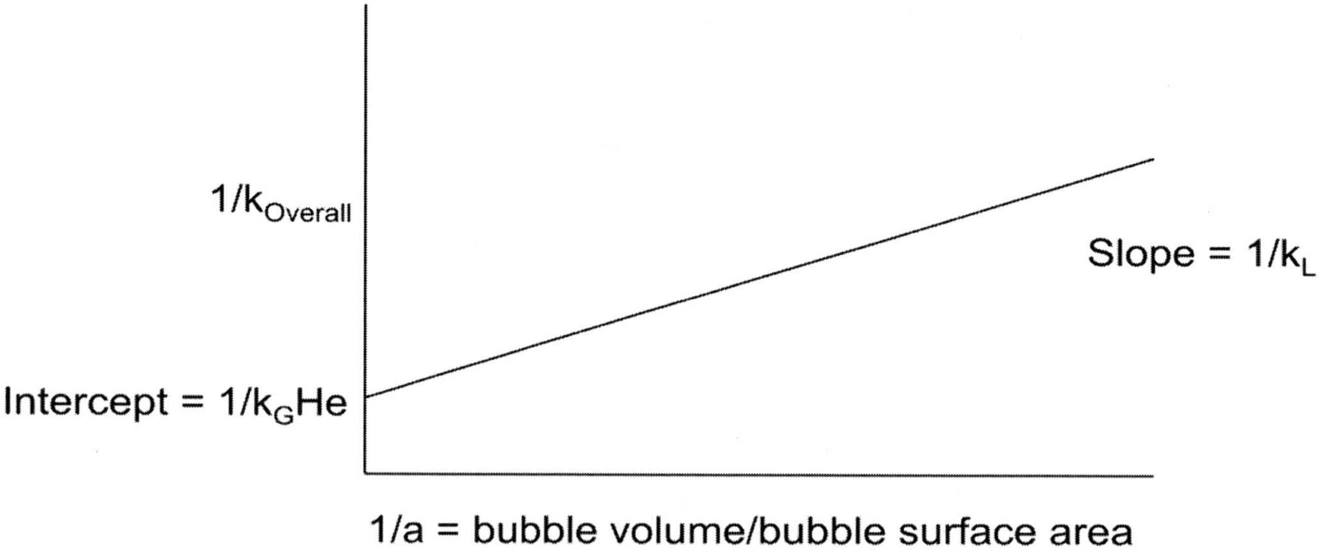

Figure 5.
Diffusion Rate and Reaction Rate Control Regimes.

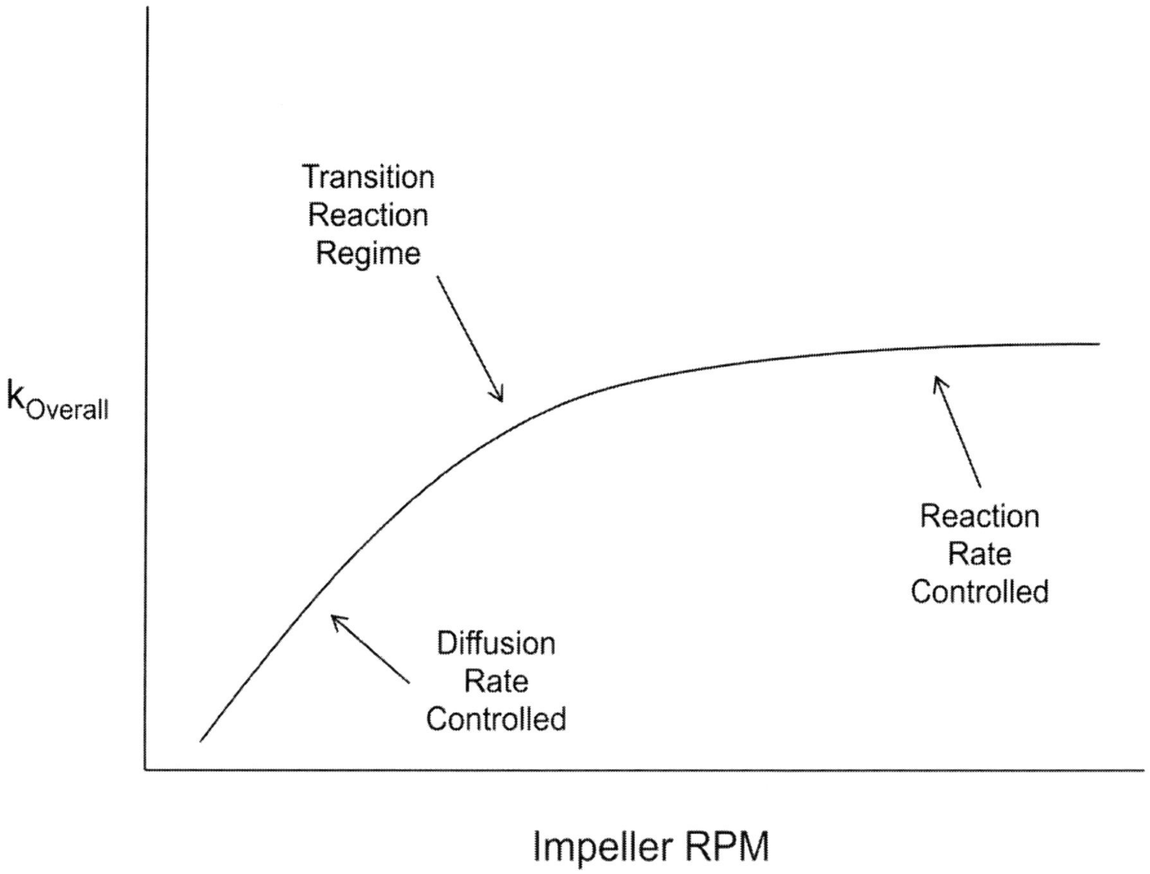

Figure 6.
Power Number as a Function of Impeller Reynolds Number.

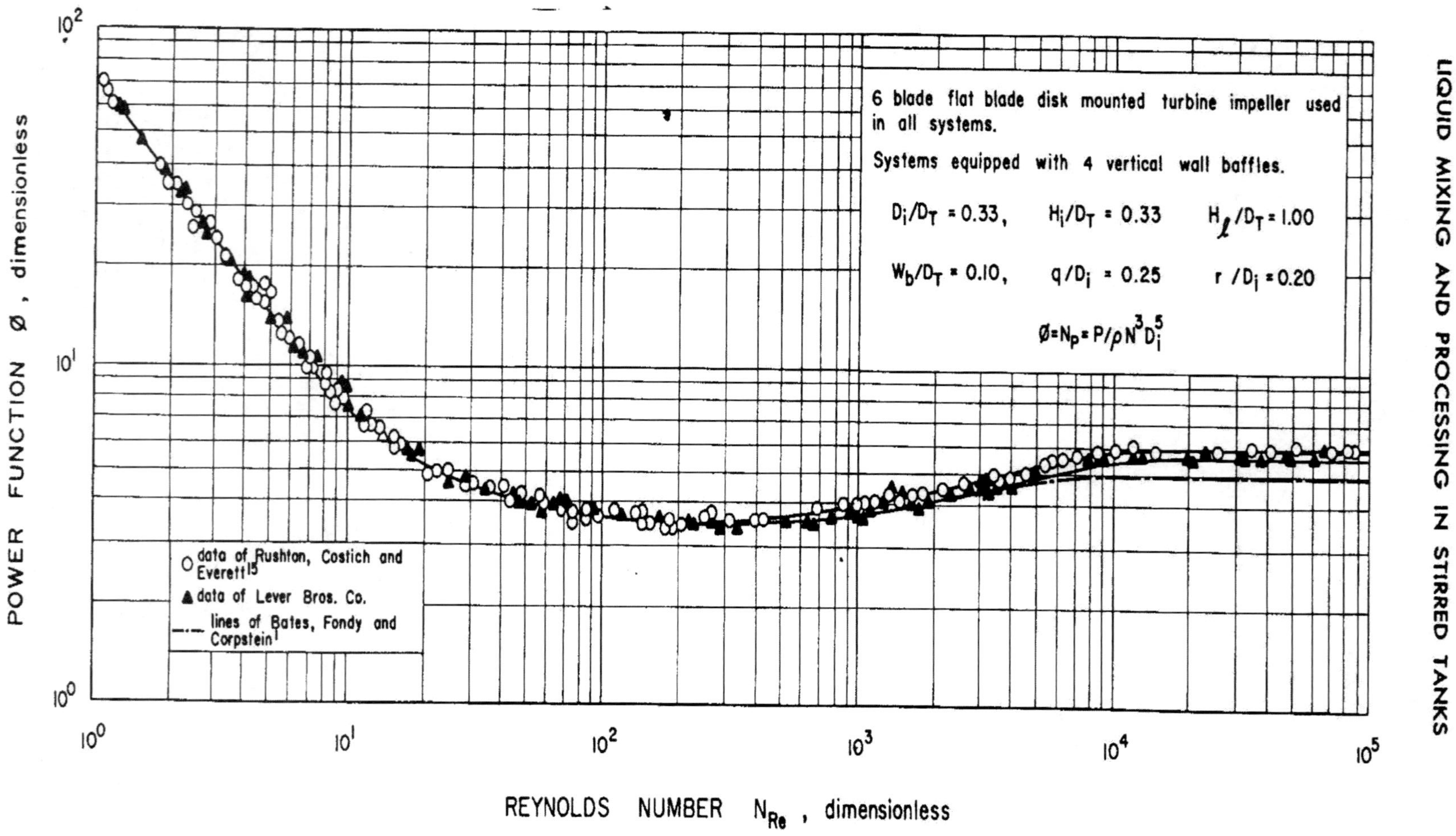

Figure 7.
Power Number and Power Function as a Function of Baffle Placement and Size in a Semi-batch Reactor.

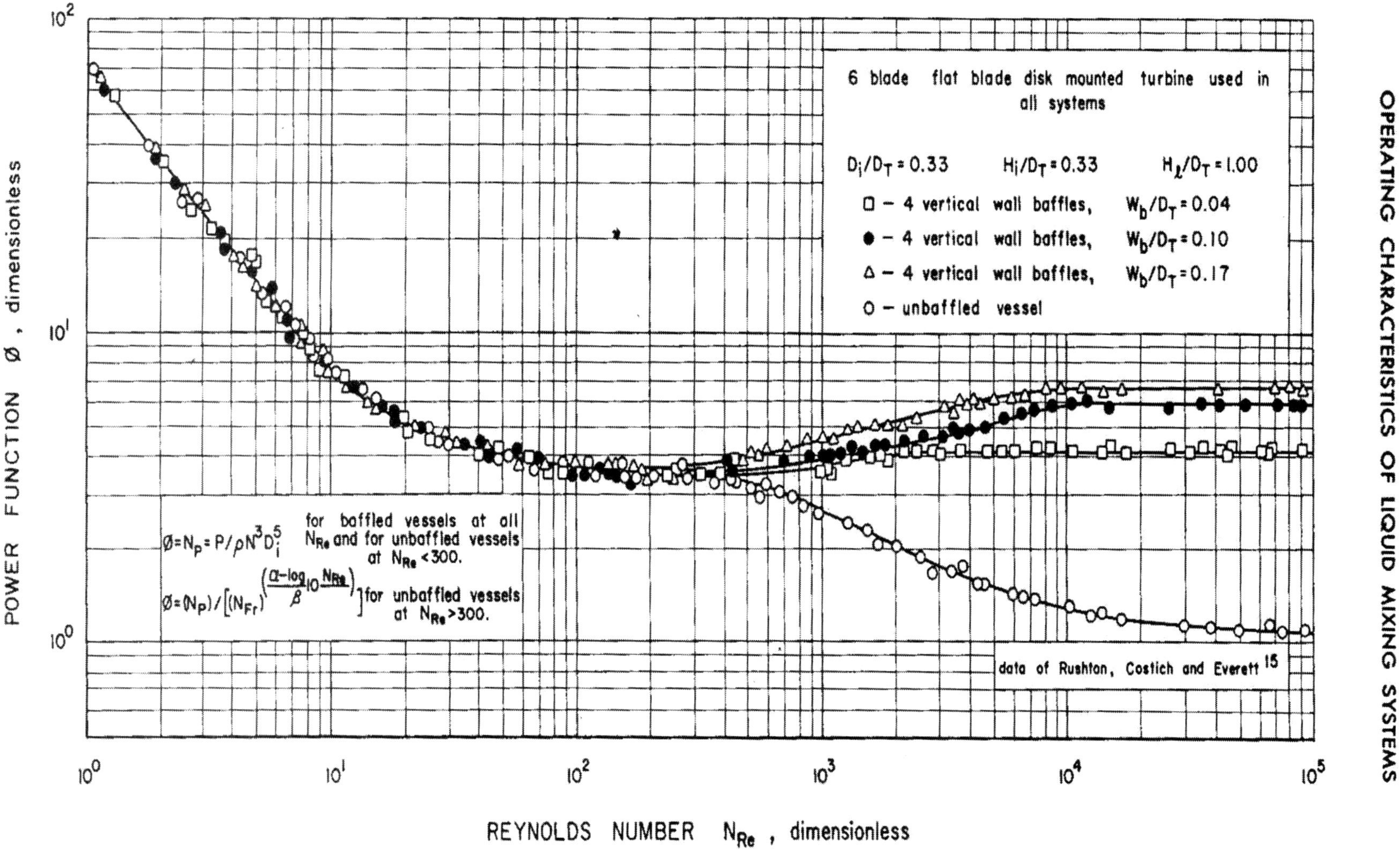

Figure 8.
Power Number as a Function of Agitator Configuration.

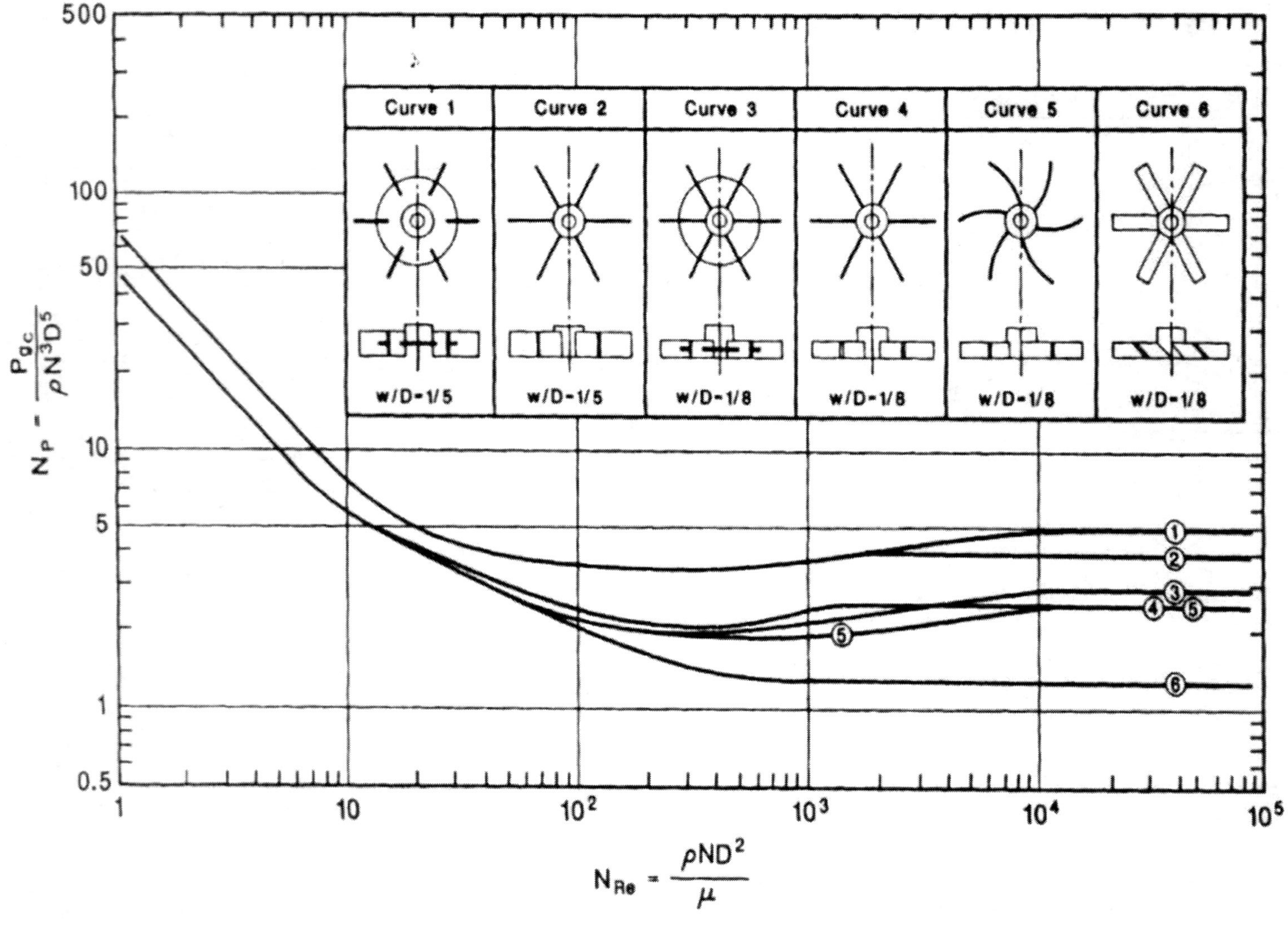

(149d) On Stability, Dynamics of Lunar Oxygen Production From Ilmenite in a 0.167 G Fluidized Bed

Kal Renganathan Sharma, Department of Chemical Engineering, Prairie View A & M University, Prairie View, TX and Ramlingam Radhakrishnan, Civil & Environmental Engineering, Prairie View A & M University, Prairie View, TX

Main Findings

Ilmenite ($FeTiO_3$) can be found in the lunar crust, in beach sands in Canada, USA and Australia. Oxygen can be generated from these rocks for use in lunar missions and lunar colonization by reduction reactions. Other materials sucah as $CaAl_2Si_2O_8$may also be considered (Caroll, 1983). Both ilmenite and anorhite are lunar-surface-minable, occurs in soils, breccias, rocks and basalts. Hydrogen appears as both reactant and product in the electrolytic reactions. This can be modelled usind the***Reactions in Circle***scheme (Sharma, 2010). For certain values of reaction rate constants subcritical damped oscillations in concentration of Oxygen can be expected. These can be performed in a 0.167G lunar fluidized bed. The bubbling bed model is revisited in light of the microgravity considerations. Transient dynamics of the ilmenite reductions reactions in a CSTR were evaluated. The system was found to be stable for small values of Damkohler numbers for forward reduction reaction, reverse reaction and electrloytic split of water reactions. For larger values of Damkohler numbers such that $(-1 + Da_2 + Da_b + Da_f) > 0$ one of the species concentration would be stable and the other would be unstable. When the "b^2-4ac" in the quadratic auxilliary equation is a perfect square, oscillations will not arise. Studies of transient dynamics in a PFR were undertaken. The resulting equation in dimensionless form was found to be of the hyperbolic PDE of the second order. The solution was obtained using the method of separation of variables. Significance of wave form of solutions are discussed. Incentives for lunar oxygen production from lunar materials as opposed to supplying it from earth have been outlined in Criswell (1983) and Davis (1983). Thermodynamic and scale-up considerations are discussed.

Background

Scientific and industrial projects on the moon is less science fiction and more a real possibility. A small batch of concrete was synthesized using 1.4 ounces of lunar soil provided by NASA at Stokie, IL by Construction Technology Labs Inc. Side by side comparisons on compressive strength, static and dynamic modulus of elasticity, Poisson's ratio at peak load, modulus of rupture and coefficient of thermal expansion between concrete made with lunar soil and that made with earth soild was performed. The properties of the specimen prepared with lunar soil were found to be as good as the one prepared with earth's soil. Water is not available in abundance in moon despite the reports from Chandrayan (2009). Moon Minearlogy Mapper, M^3on Chandrayan-1 has detected water on the moon. M^3detected absorption features near 2.8 - 3.0 µm on the surface of the moon. Retention of OH and H_2O is an ongoing surficial process. Water can be created by heating ilmenite with hydrogen. The normal boiling point of liquid hydrogen is -252.8^{0}C. Concrete is more suitable for lunar construction compared with other materials. Production of concrete requires only 1.3% of energy needed to make steel. This may reflect in cost savings. Minimal loss of concrete strength at lunar temperatures that vary between -250 F to + 250 F facing the sun. Cement can make excellent gamma ray absorbing materials. Concrete is also abrasion resistant besides.

(149e) Effect of Powder Properties and NIR Illumination Spots on the Precision and Exactitude of Multivariate Calibration Models for NIR Spectroscopy

Luis Guillermo Obregon-Quiñones, Chemical Engineering, University of Puerto Rico, Mayagüez Campus, Mayagüez, PR and Leonel Eduardo Quinones- Fontalvo, Chemical Engineering, University of Puerto Rico - Mayaguez Campus, Mayaguez, PR

During the last decade the study and implementation of NIR spectroscopy has increased exponentially giving a valuable advance to the science especially in the study of powder behavior. The pharmaceutical industry is moving toward the use of continuous production enabling the use of continuous powder mixing. This last step involves the use of important sensors to measure homogeneity of mixing online which has been a great challenge. NIR spectrometer plays an important role in this matter and that is why a continuous deep study is necessary in order to be completely sure of the degree of prediction that can be achieved under different physical conditions. Considering this subject different calibration models were made in a complete range of concentration from 0 to 100% w/w for Naproxen as API. The excipient used was Lactose Monohydrate with Magnesium Stearate 1%. The mixtures were prepared with intervals of 1 % from 0 to 30% and with intervals of 5 % from 30% to 100%. These powders have different properties making the mixture a challenge for the NIR. The powder mixing was done initially in a V-blender and the mixing time for all the mixtures were the same. Then the powders were introducing in the continuous mixer and NIR spectra were taken with two different NIR spectrophotometers at the output stream and samples were taken to be analyzed and compared with UV spectrophotometer. Precision and exactitude results had variability that depends on the cohesion of the mixtures, average particle size and type of NIR.

(149f) On Ultra High Performance Nanostructured Concrete by Polymer Impregnation

Kal Renganathan Sharma, Department of Chemical Engineering, Prairie View A & M University, Prairie View, TX and Ramlingam Radhakrishnan, Civil & Environmental Engineering, Prairie View A & M University, Prairie View, TX

There are 200 million automobiles and other vehicles on the road. Federal highway infrastructure needs to be rebuilt. The interstate highways were built in the 1950s under President Dwight E. Eisenhower. President Barack Obama has called for extensive infrastructure building in the transportantion sector under his stimulus and jobs plan. Concrete with compressive strength greater than 207 MPa (30,000 psi) needs to be developed. Nanostructured concrete may be a way to attain this goal. President John F. Kennedy presented his vision to send a man to the moon by the ned of the decade and bring him back safely during his presidency between 1961-1963. Lunar concrete is another goal.

Concrete is the most used material by mankind next to water. About 7.5 $km^3$of concrete is prepared every year. This is roughly 1 $m^3$for every man, woman and child on the face of this planet, earth. It is a 35 billion dollar industry employing about 2 million workers in United States of America. Over 55,00 miles of highways are paved with reinforced concrete and pre-stressed concrete. Concrete is used in road and bridge construction, for support of elevated road beds and highways, pilings, pillars, etc. They are also used in skyscrappers, hgh rises, etc.

The next generation concrete is expected to: (i) offer a service life of over 100 years; (ii) possess compressive strength upto 207 kPa; (iii) require less construction time; (iv) by watertight; (v) crack-free; (vi) versatile; (vii) durable; (viii) corrosion resistant; (ix) abrasive; (x) resilience to impact; (xi) engineering properties comparable to stainless steel, cast iron and ceramics. UHPC, ultra-high-performance-cement can be used to make thinner sections, larger spans and taller high rises. Formulation to make cement composites include portland cement, silica fume, quartz fluor, fine silica sand, high range waste reducer and steel or organic fibers.

The methods of preparation of concrete has changed since the days of the Roman empire when ***opus caementicium***or cement was made from quicklime, pumice and pozzolan. J. Aspdin in 1824 invented portland cement. Raw mix is prepared this time and age followed by the clinker product and cement. Raw materials in the form of slurry are made from CaO, calcium oxide, SiO_2, silicon oxide, Al_2O_3, alumina, Fe_2O_3, ferric oxide and MgO, magnesium oxide. Quarried from local rock, they are ground together in a raw mill. Milled down to less than 100 μm the raw mixture is heated in a cement kiln with temperatures of 1400 - 1450^{0}C. A complex set of chemical reactions leads to a sintered product. The enthalpy of formation of clinker from $CaCO_3$and clay minerals is 1700 $kJ.Kg^{-1}$. Emergence of nanotechnology (Sharma, 2010) has enhanced the expectations for: (i) stronger concrete; (ii) development of super plasticizer; (iii) reduction of clinkering temperature; (iv) cold sintering of clinker materials in mechano-chemical reactions; (v) nanocatalysts for quicker hydration reactions; (vi) grinding aids for superfine clinkers; (vii) binders reinforced with nanotubes and nanorods; (viii) biomimetic materials; (ix) cement based composites; (x) step change improvement in ductility and toughness; (xi) reduced micro-cracking; (xii) self-healing materials using fullerenes. The use of an effective super plasticizer can effect dispersion stability in***nanosilica***products and thereby improve the strength of Portland cement mortars at all ages of hardening. Longevity of concrete can be improved by introducing reinforcing carbon fibers. These fibers can come from bituminous coals (Renganathan and Zondlo, 1988). Improved flexural, deflection and ductile properties of conrete can be effected by proper introduction of carbon fibers. Voids in concrete may be filled usinng nanofibers of carbon. Void fillage can result in reduction of crack formation as energy gets deflected and dissipated during propagation.

39 different nanostructuring methods that may be used in preparation of nanostructured concrete are reviewed. These methods are as follows: ***(1) sputtering of molecular ions;(2) gas evaporation; (3) process to make ultrafine magnetic magnetic powder; (4) triangulation and formation of nanoprisms by light irradiation ; (5) nanorod production using condensed phase synthesis method; subtractive methods such as; (6) lithography; (7) etching; (8) galvanic fabrication; (9) lift-off process for IC circuit fabrication; (10) nanotips and nanorods formation by CMOS process; (11) patterning Iridium Oxide nanostructures; (12) dip pen lithography; (13) SAM, self assembled monolayers; (14) hot embossing; (15) nanoimprint lithography; (16) electron beam lithography; (17) dry etching; (18) reactive ion etching; (19) quantum dots and thin films generation by; (20) sol gel; (21) solid state precipitation; (22) molecular beam epitaxy; (23) chemical vapor deposition; (24) CVD; (25) lithography; (26) nucleation and growth; (27) thin film formation from surface instabilities; (28) thin film formation by spin coating;(29) cryogenic milling for preparation of 100-300 nm sized titanium; (30) atomic lithography methods to generated structures less than 50 nm; (31) electrode position method to prepare nanocomposite; (32) plasma compaction methods; (33) direct write lithography; (34) nanofluids by dispersion. Thermodynamic miscibility of nanocomposites can be calculated from the free energy of mixing. The four thermodynamically stable forms of Carbon are diamond, graphite, C_{60}, Buckminster Fullerene and Carbon Nanotube. 5 different methods of preparation of CNTs, carbon nanotubes are discussed.***

Polymer impregnated concrete specimens can improve fuel efficiency of automobiles, i.e., mpg. 20 different block copolymer precursor morphologies can lead to 20 different nanostructures post structuring. The structures are a spectrum ranging from layers to orderly spherical dispersions. Streaks may also be found. It is not clear why certain morpholgies lead to better performance. Thermodynamic considerations of miscibility and phase separation are also pursued. Minimum allowable size estimates based on Gibbs free energy and surface tension need to evaluated for agglomerative stability.

A concrete specimen can be designed as follows: (i) PAn, polyacrylonitrile fibers are formed by electrospinning. Nanofiber formation can be better controlled. Prototypical concrete comprises of 15% cement, 18% water, 8% air, 28% fine aggregate and 31% coarse aggregate. PAN fibers and nylon fibers can be impregnated into the cement composite. Silicate platelets may be used with the intergallery distance in the nanodimensions (< 100 nm). The bond between fibers and the cementicious portion may be strengthened by ***pozzalanic reations***. Slakes lime, $Ca(OH)_2$, calcium oxide may be reacted with amorphous siliceous materials to form calcium silicate hydrate, CSH. The general formula for calcium silicate hydrate ($CaH_2SiO_4.2H_2O$). Consequence of reaction products may be ***swelling***. Interfacial adhesion between the fiber and the cementicious portion is an important consideration. Interface changes with time as teh cement matures or undergoes time dependent volume changes. Optimal volume fraction of nanofibers for better performance of nanocrete can be arrived at from mathematical modelling. Current practioners use 0.5% PP, polypropylene fibers.

Cost reductions in nanostructuring are proposed. The cost has come down from $200,000 per pound in the Arizona process to $200 per pound in the second generation combsution synthesis process. HeIM, helium ion microscopy or gamma ray characterization are proposed to study the nanostructured speciment. The Raleigh resolution criterion is 200 nm. So optical devices are not sufficient.

The fundamental structure of cement is not clear. A new molecular model for cement hydrate was unveiled at MIT, (Pellenq. et. al., 2009). Previous models based on natural mineral analogs indicated that cement nanoparticles consist of long SiO_2 chains interspersed with neat layers of CaO and H_2O. The new model based on neutron scattering data is a hybrid of crystalline and amorphous $CaO-SiO_2-H_2O$. Hybrid model contains shorter chains of SiO_2 with flaws that undercut the CaO layers leading to voids that are filled by water molecules. Minor disorder provides allowance for cement to stretch or compress rather than snapping. With molecular models similar to this cement chemical structure can be manipulated in order to obtain the desired strength of the concrete, durabilioty and environmental qualities such as low CO emissions.

AUTHOR INDEX

AUTHOR INDEX